ARABS, OIL AND ENERGY

ARABS, OIL AND ENERGY

EDGAR C. JAMES

MOODY PRESS
CHICAGO

pu
Library of Congress Cataloging in Publication Data

James, Edgar C.
  Arabs, oil, and energy.

  Includes bibliographical references.
    1.   Arab countries—Politics and government.
  2.   Petroleum industry and trade—Arab countries.
  I.   Title.
  DS63.1.J35   309.1'17'4927   77-26112
  ISBN 0-8024-0294-1

Printed in the United States of America

DEDICATION

To my wife, Barbara,
whose life and love
have given me joy and strength

Contents

Acknowledgement

Some of the material in this book first appeared in my articles "Prophecy and the Common Market," *Moody Monthly*, March 1974; "Here Come the Arabs," *Moody Monthly*, July-August 1974; "Will We Have Peace?" *Moody Monthly*, March 1976; and "Down Payment on Peace in Israel," *Moody Monthly*, April 1976.

Preface

"Oil! Oil! Oil!" This is the cry of our world. We want oil for heating, for travel, for industry. We need it for jobs, for economy, for world survival. But will it be said that the industrial revolution, which began in the nineteenth century, will end in the twentieth because of a lack of oil?

The Arab nations control most of the world's oil supply. Through embargoes, diplomacy, and cartels they have cornered the world market and price. In many ways they have us over a barrel.

What is going to happen to this oil situation and the nations that control it? And most important, what is going to happen to our world? Will we be able to survive?

Some may seek answers to these problems through astrology or spiritism. Others may take a "come what may" attitude. Still others may bury their heads in the sand like the ostrich. But wouldn't you like to know what is going to happen in the future?

The real answers are in a Book written almost two thousand years ago. The solutions it gives to other problems have proved it to be absolutely true. It makes sense to examine it for answers to these problems.

The pages that follow show my search for these answers and my conclusions. Where important, I have included documentation but have tried to keep notes to a minimum so as not to slow down the reader. So get ready for an exciting journey as we examine the Arab revival and world survival!

1

Arabs, Oil, and the Twentieth Century

"The Arabs are coming! The Arabs are coming!"

If we had said that a few years ago, nobody would have believed us. But since the Arab oil embargo, the world has drastically changed. Like a giant Rip Van Winkle, nations that were asleep for many years have awakened with a start. Today the Arabs are not only coming, they are here!

Take the oil-rich Arabian Gulf country of Kuwait, for example. A short time ago this country was a land of primitive, mud-walled, subsistence villages in which no one was interested. All of that is changed. Its people enjoy full and free education up through the Ph.D. degree, free medical care, modern housing, easily affordable servants, big cars, stereos, and color TVs. There are no Western worries such as income taxes, hospital bills, telephone bills, or unemployment. What's more, at a time when many nations' economies are

11

This extension of Ju'aymah terminal's offshore flow control and metering platform demonstrates the mushrooming Middle East oil business. The terminal's nominal shipping capacity is three million barrels per day. Photo courtesy of Aramco.

standing still or declining, this nation's income multiplies year after year!

The world today is witnessing the greatest redistribution of wealth in history. A handful of Arab nations, emerging from centuries of deprivation, find their fortunes soaring on the galloping price of oil. Overnight, the Arab states have become a major economic, political, and military bloc. Just how much money and power they accumulate and how they use it will determine the future course of the world.

For many years when the eyes of the world turned to the Middle East, they focused on Israel. Recent events, however, remind us that Arabs, too, have a prophetic future. Scripture makes it clear that theirs is no passing role in history, but that from the beginning they have been destined to remain and share in end-time events.

The modern rise of the Arab nations began on May 26, 1908, when drillers first struck oil at Masjid-i-Salaman (Mosque of Solomon) in Iran, the site of an ancient temple. Because oil-skilled nations were more interested in first developing their own resources, serious oil production in the Middle East did not begin until after World War II. For instance, in 1938 the Middle East produced only 5.5 percent of the total world output of oil. Thirty-five years later it accounted for 40 percent.

"Allah was generous when he gave us the oil," a geology professor at the University of Riyadh said.[1] The results of Arab possession of that oil are having far-reaching economic, political and military effects in our world today.

13

Students of history will remember that following the fall of the Roman Empire, Arab nations were numbered among the world powers. After the defeat of the Moors at the Battle of Tours in A.D. 732, the Arab states dropped back to places of relative obscurity. Many have written them off as no longer significant factors, but the Bible indicates otherwise. In fact, the prophet Isaiah in foretelling the future reign of Christ spoke of a day when the Lord will say, "Blessed be Egypt my people, and Assyria the work of my hands, and Israel mine inheritance" (Isaiah 19:25).

The twentieth-century world will not soon forget the Arab's latest rise to prominence. This came at the height of the Arab-Israeli Yom Kippur War in October 1973, with an announcement of an oil embargo and an increase in oil prices.

The results in this country in terms of waiting lines at filling stations, trucking tie-ups and lowered thermostats are only too well known. In Western Europe, some countries imposed gas rationing and banned Sunday driving. Most seriously affected was Japan, which imports nearly all its oil.

What began as a sudden change has become a whole new way of life. Cheap gasoline was no more, and what there was had to be used sparingly. Big cars were scaled down and gas mileage went up. Lower speed limits were made permanent. Heating oil became scarce, and natural gas redistributed. "Conserve energy" were the national watchwords, as fuel bills multiplied.

Meanwhile, the flow of oil money into Arab lands

14

Sheik Yamani is Oil Minister of Saudi Arabia. Photo courtesy of Aramco.

became a flood. Middle East oil revenue multiplied five times within a year after the embargo. In a fanciful bit of arithmetic, London's *Economist* calculated that members of the Organization of Petroleum Exporting Countries were piling up reserves fast enough to buy the Bank of America in 6 days, IBM in 143 days and all of the companies listed on the world's major stock exchanges in 15½ years. Without question, the Arab nations had become a major power bloc overnight.

Once again a contemporary development seems to have been in view in the Scriptures. Referring to the Arab peoples, the Bible speaks not only of judgment but also of blessing. The references are extensive. Whole sections of Isaiah, Jeremiah and Ezekiel are concerned with Arab peoples. More than 140 references are made to Assyria, more than 730 to Egypt, and some 180 to Moab during various periods of the past and future. Books like Obadiah and Nahum are wholly concerned with Arab nations and Isaiah 19 begins with the words, "The burden of Egypt."

Why is oil such an important commodity? It is a necessary and useful resource, and it is unevenly distributed throughout the world. Noah knew something of oil's value; the Bible records he used pitch to caulk the ark (Genesis 6:14). Archaeologists have found that the walls of Babylon were cemented with bitumen. Ancient physicians collected oil and used it for medicinal purposes.

It was not until the nineteenth century that oil was extracted in any quantity. It was then that the Scots learned how to make kerosene from coal. Kerosene quickly replaced other lamp fuels since it was better and cheaper. Encouraged by this success, companies investigated crude oil as another possible lamp oil source.

And then it happened. Edwin L. Drake, after continued persistence, struck oil with a drilling rig on August 27, 1859. The petroleum gushed forth like a geyser, and a new industry was born.

By 1900, America had a booming oil industry. In

1902, Spindletop, the first oil well near Beaumont, Texas, erupted. In 1908, oil was first discovered in the Middle East.

Today, oil is used for all types of transportation and industry. Oil and natural gas provide three-fourths of all the energy for the United States. Although there are alternative sources of energy, none is as economical or available as oil. Roger Sant, once the Assistant Federal Energy Administrator, said, "The more I look at the alternative resources to cheap oil and gas, the more appalled I am at the cost."[2]

The Middle East contains more than half of the world's known oil reserves (as compared with the United States, which, with Alaska, has only 5 percent). The advantages of Middle East oil are that it is plentiful and can be produced cheaply since it is close to the surface and does not require extensive drilling. It also lies close to the sea, so that pipeline and other costs are low. The quality of the oil is excellent because it is low in sulphur, and therefore does not present environmental hazards.

Whoever controls this oil controls the destiny of the nations that need the oil. Industrialized countries must have oil to fuel their industries and to feed their military machines to protect themselves. They are willing to pay dearly for their oil and may one day even be willing to fight for it.

The threat of future fighting for oil caused President Carter to propose an energy program within his first hundred days of taking office. Before the American

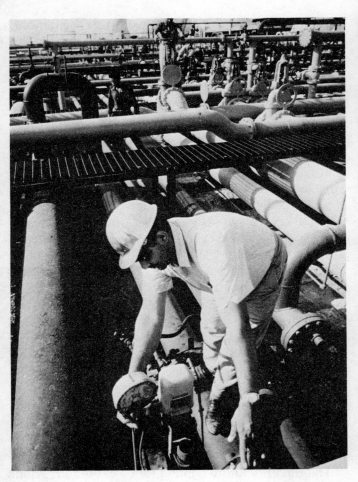
Workman inspects and tests oil plants and equipment in Saudi Arabia. Photo courtesy of Aramco.

people and a joint session of Congress he pronounced "the moral equivalent of war" to save the nation from economic, social, and political crisis. "The most important thing about these proposals is that the alternative may be a national catastrophe," Carter said. "With the exception of preventing war, this is the greatest challenge our country will face during our lifetime."[3]

Even with energy legislation and conservation, however, vast imports continue. The increase in the rate of consumption has slowed, but as population and living standards increase, consumption of energy does not decrease.

What frightens economists is that the wealth of the energy users is being transferred to the energy producers. What frightens politicians is that the energy of small producing nations may soon be used up. This means that the energy users will be at the mercy of a very few, large, energy-producing nations. What frightens military experts is that Carter's warning of national catastrophe may become a reality!

What are the Arab nations doing with all their money? Some is being used for economic progress. For instance, Iraq is spending one billion dollars for an agricultural program to increase arable land, water storage, and food production. Saudi Arabia has several development projects on the drawing board. In the Arabian Gulf countries, entire cities are being built where desert sands once reigned.

But progress takes time as well as money for a people who live much as they did over two thousand years ago.

This is why a lot of funds are invested in Europe and the United States. Kuwait, for instance, owns interest in a German automobile corporation, a British real estate company, and a skyscraper in Paris. The United Arab Emirate ambassador to London owns the famous Mereworth Castle. The Arab country of Abu Dhabi owns a substantial interest in London's Commercial Union Assurance, and a Saudi Arabian bought control of Detroit's Bank of the Commonwealth, the Detroit area's fourth largest bank.

"A bank makes sense because we have the material of banking—capital," explained Roger Tamraz, a Harvard-educated Lebanese investment banker. But there was another factor, too. "We wanted Arab capital

Tankers line the loading piers of Ras Tanura terminal, Saudi Arabia. Photo courtesy of Aramco.

to look responsible," said Tamraz.[4] Arab capital extends the Arab influence throughout the world.

Arab capital is also being used for political purposes. Non-oil-producing Arab countries such as Egypt and Syria have received large contributions to help build Arab unity. Prince Fahd Ibn Abdel-Aziz, heir to the Saudi Arabian throne, lists his nation's priorities this way: "Our first duty is to the inhabitants of our kingdom. Then we will take care of the Arab or Muslim countries which are our long-standing neighbors and friends. Then we will help the developing countries without distinction. They deserve our aid more than the rich countries."[5]

The Arabs have also used their funds for military purposes; much of their money has gone for arms more than alms. The Gulf countries all have armies. Iran, Saudi Arabia, and Kuwait have spent billions for arms, for their own protection as well as the rearmament of Egypt. The Middle East has become a military fortress.

The Arab nations have rapidly become an economic, political, and military power bloc that is swiftly changing our world. Not many know about their past, and few know anything about their future. Who are these people who have suddenly come to life and are seeking to change the course of world history?

2

Who Are the Arabs?

In the bullring of Madrid, Spain, tension builds as every eye is riveted to the brilliantly-dressed matador. He bravely fights the savage bull, then kills him with a glinting sword. "Olé! Olé!" roars the Spanish crowd, a cry that once was "Allah!"

"Khiva [Russia] was long a center of Arab learning. Its most famous son was al-Khwarizmi."[6] Al-Khwarizmi wrote treatises on mathematics, and the word *algebra* was his invention. His original name was a long one that contained the Arabic system of numerals, one through nine plus zero, that the world uses today.

An Arab born near Tehran, Iran, was "the greatest and most original of all the Muslim physicians."[7] His name was al-Razi, and he invented the seton used in surgery. He wrote treatises on smallpox and measles. His portrait adorns the great hall of the School of Medicine at the University of Paris.

Although Spain, Russia and Iran are far removed from the Arab countries, they are much involved with

Arab culture today. This is because there once was an Arab Empire that stretched throughout Europe, Africa, Asia, and some believe, even to China. History records that this empire passed the torch that helped light Europe's Dark Ages.

"You will find more than four thousand Arabic words still common in Spanish," said a professor at the University of Granada, Spain. And even in English we have such Arab words as *azimuth, zenith, alkali, alcohol,* and *cipher.* Today Arab mosques, a symbol of this previous empire, are seen throughout the world in Spain, Russia, Iran, India, and on the Jewish temple site in Jerusalem.

Who are these people who may again rule the Middle East, this strategic part of the world? A popular concept is that the Arabs descended from Ishmael, the son of Abraham and the Egyptian slave, Hagar. Like Jacob, Ishmael had twelve sons who became heads of tribes (Genesis 25:16).

Muhammad, the seventh-century prophet of the Islamic religion, claimed descent from Ishmael and therefore Abraham. The Muslim shrine, the Dome of the Rock, is on the Jewish temple site in Jerusalem and is built over the rock where Muslims believe Abraham offered his son (Genesis 22).

But the term *Arab* applies to people other than the physical descendants of Ishmael. Today the term is used to designate all the inhabitants of the Middle East and North Africa who speak the Arabic language and identify with Arab culture. Arabs constitute the major proportion of at least sixteen countries with a population of more than 130 million people, of whom 116 million are Arabs or live much like the Arabs.

Arabs may look or dress differently from each other, depending on their background and location. For instance, in Saudi Arabia, one might find an Arab looking much like an Old Testament patriarch, with flowing robes and beard. Beirut, in contrast, may be the place one might find a decisive, blonde career woman or an executive officer of a bank. In Tel Aviv, there may be an Israeli Arab who speaks four languages, has a degree from the American University of Beirut, and can trace his line through many generations of living in Jerusalem. In charge of the oil fields in Kuwait or Saudi Arabia, there may be an Arab with graduate degrees from Stanford or Harvard.

Several nations considered Arab existed before Ishmael was ever born. For instance, Egypt, Syria and Assyria (present-day Iraq) are mentioned in the table of nations (Egypt as Mizraim, Genesis 10:6; Syria as Aram and Assyria as Asshur, Genesis 10:22). The Edomites were descendants of Esau, and the Moabites and Ammonites were descendants of Lot's sons by his two daughters.

Although the Edomites, Moabites, and Ammonites did not originate with Ishmael, they were related to him. Ishmael's mother, Hagar, was an Egyptian. Others are related through Shem, the son of Noah. Even Abraham is called a Syrian (Deuteronomy 26:5). This has caused some to believe that the promises made to Ishmael may extend to the whole Arabian family.

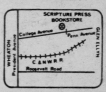

SUBURBAN TRIB Mon., Oct. 9, 1978   II   **23**

Wesleyan Seminary, the Garrett Biblical Institute, and the Gammon Theological Seminary in Atlanta. He had established the Gammon Seminary specifically for the education of Negroes entering the ministry.

In honor of the contribution Gammon made to Batavia, Marconi has named his new business "Gammon Corners."

In a $70,000 renovation, he has restored the exterior to an approximation of its original blue, and has painted the trim cream and maroon, a treatment popular in the 1800s. Ms. Donali, Ltd., occupies the first floor of the building, selling moderately and higher priced women's clothing. Josie's, across the street, will sell moderately to "upper" mod-

The holy mosque is located at Mecca. Photo courtesy of Aramco.

Many of the early prophecies concerning the Arab peoples seem to have been fulfilled over a period of time. For instance, God promised that from Ishmael would come a great nation (Genesis 17:20). This has come to pass, whether the prophecy refers to Arabia or the Arab peoples as a whole. The prediction that Ishmael would be a "wild man" (Genesis 16:12) seems to foreshadow the roaming life followed by so many of the Arab nomads. Today five million Arab bedouins live in tribes and have not changed their customs in over two thousand years.

The prediction may also picture the disunity so evident in the Arab world. Arabs have spent more time fighting each other than in finding avenues for cooperation. There was a spark towards unity in the 1860's when the Syrian Scientific Society kindled pride in Arab history, literature, and culture. But Arab nationalism has not led to Arab unity.

For instance, in 1961, the attempted Iraqi take-over of Kuwait was prevented by British troops. In 1962, Egypt intervened for 5½ years in a civil war in Yemen. From 1971 to 1973, Syria had no diplomatic relations with Jordan. In 1976, Syrian troops took over the country of Lebanon. In addition, Oman and South Yemen have been in a continual state of war because South Yemen supports rebels in Oman.

The Scriptures also declare that Ishmael and his descendants would "dwell in the presence of all his brethren" (Genesis 16:12), that is, his descendants would live adjacent to other descendants of Abraham. Though

Israel has been removed from her land three times, descendants of Ishmael have remained in relatively their same locations. Even today, the nations that surround Israel are Arab.

What important contributions have Arab nations made to biblical history? An examination of Scripture shows that they turned to their own gods and did not follow the Lord, so they oppressed Israel. Egypt was at first a haven for the sons of Jacob when famine struck the land of Palestine. Then Egypt enslaved the people for four hundred years and released the Israelites only when plagues were sent against them. Later, the nation of Assyria came against Israel and took many of the people captive to Babylon. However, both Egypt and Assyria are subjects of important biblical teaching that shows not only God's judgment of them, but also God's future blessing of them.

When did the Arab Empire begin? During Bible times, the Arab countries were seen as separate nations. A few centuries after the Bible was completed, the Arab world was fused into one vast empire. The catalyst for this fusion was a man named Muhammad. His name means "highly praised." He was born in Arabia in A.D. 571 in a tribe that kept a special shrine in the city of Mecca. Although little is known of his early life, it is known that at the age of twenty-five he married a wealthy widow named Khadijah, who was fifteen years his senior. This arrangement gave him time for seclusion and contemplation. One day, in the midst of meditating, he thought he heard the voice of Gabriel calling him to the prophetic office. He wrote the revelation he received in the *Koran,* the holy book of the Muslim world.

Muhammad gained few converts at first. Some from the slave and lower classes believed, but others had to be persecuted into believing. Eventually, he was invited to make the city of Medina in Arabia his home. He arrived there on September 24, A.D. 622. This date is important because this is when the famous Hegira took place and is the official starting-point of the Muslim era. This marked the turning point in Muhammad's life. He had left Mecca, the city of his birth, as a despised prophet, and he had entered Medina, the city of his adoption, as an honored chief. In addition, Islam, which had been on the defensive, now turned to the offensive. It not only became the state religion, but in Medina, it was the state itself.

The prophet broke with Judaism and Christianity in faith and in forms of worship. For instance, Friday was considered the holy day instead of Saturday or Sunday. The call from the minaret replaced that of trumpets and bells, and prayer was directed toward Mecca, not Jerusalem. Ramadan was a fixed month of fasting, and a pilgrimage to Mecca was authorized.

The climax of Muhammad's work was in A.D. 628 when he led a group of 1400 Muslims to Mecca. After a great struggle, he entered the city and smashed all the idols. Delegations from near and far came to offer their allegiance to him. The whole country of Arabia recognized him, as well as people from other nations. He died unexpectedly on June 8, A.D. 632.

The Muslim shrine Dome of the Rock occupies part of the Temple site in Jerusalem. Photo courtesy of Israel Ministry of Tourism.

The little wealth Muhammad left became state property. He had about a dozen wives and a number of children, but only one, a daughter, survived. He also left the Koran containing laws governing fasting, prayer, alms giving, social and political ordinances dealing with marriage and divorce, the treatment of slaves, prisoners of war, and enemies. These laws especially favored slaves, orphans, the weak, and oppressed.

29

It was in his farewell sermon that Muhammad insti-
tuted Arab nationalism, a brotherhood of faith rather
than tribal kinship. He admonished, "Oh ye men! Har-
ken unto my words and take ye them to heart! Know ye
that every Muslim is a brother to every other Muslim,
and that ye are now one brotherhood. It is not legitimate
for any of you, therefore, to appropriate unto himself
anything that belongs to his brother unless it is willingly
given him by that brother." Although Muhammad was
an unschooled man, he was responsible for the Koran,
which is still considered by one-seventh of mankind as
the embodiment of all science, wisdom, and theology.

Islam spread throughout civilization overnight.
Within a hundred years after Muhammad's death,
Syria, Egypt, India, and Spain were conquered. The
Persian Empire was demolished, and the Byzantine
power was shaken to its very foundation.

How did such a feeble, tribal country like Arabia
conquer so much of the world? One reason was the
stimulus of Islam that awoke the East. The East then
reasserted itself after a millennium of Western domina-
tion. Another reason was a new military technique, the
use of war camels. Camels were especially useful in the
open places of western Asia and northern Africa.

Perhaps the greatest reason, however, was an
economic one. The Muslims extracted tribute from their
conquered foes and were able to enjoy many comforts
and luxuries. Muslim historians describe a reception for
envoys in Byzantium in which 700 chamberlains, 7,000
eunuchs, 160,000 cavalry men and footmen, and a

parade of 100 lions were seen. The palace was hung with gilt curtains and 22,000 rugs. In addition, there was an artificial tree of gold and silver containing mechanical birds chirping metallic songs.

What about the Arab military conquests? When Muhammad died, his father-in-law, abu-Bakr, assumed the leadership of Islam. He quickly conquered all of Arabia. Encouraged by this success, his generals were ready to move onward. Syria, to the north, came first and, within a couple of weeks, the Muslim general Khalid stood before the gates of Damascus. Although the city was beseiged for six months, it surrendered and later became the capital of the Islamic empire. With Syria as a base, the armies pushed to Armenia, Mesopotamia, and throughout Asia Minor. Even Iraq and Persia were not exempt, and by A.D. 643 the Arabs stood at the borders of India.

Meanwhile, another Arab general, Amr, moved westward against Egypt. Again the formula was the same — a rout, a seige, and the cry of victory: *"Allahu akbar,"* God is most great. Although the Egyptian forces far outnumbered the Arabian forces, the Arabs conquered Egypt quickly. They even built a new city that has survived as the city of Cairo. After the fall of Egypt, Amr, with characteristic swiftness, pushed westward throughout North Africa to the land of the Berbers in Tripoli.

In its first stage the Arab empire moved to Syria, Turkey, Persia, Egypt, and North Africa. Soon, however, the second stage began in which the eastern campaign moved through India and all the way to outer Mongolia. The westward branch, which had penetrated throughout North Africa all the way to ancient Carthage, now began to move north. It crossed the straits of Gibraltar (named for its Muslim conqueror) into Spain. Soon city after city fell to the Arabs until Spain was a province of the Caliphate, the supreme ruler of Islam.

Lured by the treasures of France, the conquest continued northward to the city of Tours in A.D. 732. Here the Arab army was met by Charles Martel, mayor of the palace at the Merovingian Court. He and his Frankish warriors withstood the attack, and the Arabs quietly vanished away by night. Charles was victorious. This battle marked the turning point in the military fortunes of the Arab Empire. The centennial of Muhammad's death, A.D. 732, marked the end of the military expansion of the Arab Empire.

The Arabian Empire, however, had had a large influence on the world for almost 800 years. It later collapsed because of the Catholic influence in the north, the Crusades, and its own internal problems. But the Arabs did pass the torch that helped to light Europe's Dark Ages before Arabia slipped into a long and fitful sleep.

But the sleep is over. Since the discovery of Middle East oil, these long-divided Arab peoples are again groping for unity. Could it be that the old Arab Empire was but a foreshadowing of what is yet to come? Is it possible that the future of these people is like dry bones coming back to life?

Although the Arab peoples did not originate many

ew ideas, they were the main bearers of the torch of ulture and civilization throughout the world. They took nportant ideas from the Greek and Persian cultures nd disseminated them throughout their empire. This is vhy, even today, there is a similarity between the art of ;pain, Persia, and the Arab countries.

Arabs have made important contributions, including some to the field of medicine. Photo courtesy of Aramco.

What important contributions did the Arabs make t
Western civilization? In the field of philosophy, the
translated Aristotle and other great philosophers an
distributed the translations to the world. In the area c
mathematics they introduced algebra, the zero, and th
decimal system. In the field of astronomy they contrib
uted the science of spherical trigonometry.

The Arabs also made an important contribution t
geography, because it was their religious duty to make
at least one pilgrimage to Mecca. They made importan
breakthroughs in medicine, such as noting the conta
gious character of disease, writing medical dictionaries
and building hospitals. The areas of chemistry and ag
riculture were also advanced by the Arabs. In the field o
literature, they developed the art of storytelling as evi
denced by the book *The Arabian Nights.*

In the field of art they made an unusual contribution
Their religion did not allow man or any living creature to
be pictorially portrayed. So they turned to abstract art.
They developed the Arabesque form of art. This was a
type of formalized, geometrical design, and occasionally
copied in the West when lettering was done for decora-
tive purposes.

The origin, history, and contribution of the Arab
peoples have been amazing. Their present rise to power
is just as amazing. Could their past be a foreshadowing
of their future? And what is going to happen to these
people and to the part of the world they live in?

3

Saudi Arabia — Money Center of the World?

Amid the barren land and desert sand of Saudi Arabia lies a city of curving streets and palm boulevards. The people live in ranch-type houses with low walls and open patios. Housewives shop in well-stocked super-markets. Students study at the College of Petroleum and Minerals. Children play in parks and on softball fields. An American would think he was in a California desert community.

This is the oil town of Dhahran in the Eastern Province of Saudi Arabia. It has changed a great deal since the time oil was first discovered nearby. One veteran oil man sums up his feelings when he first came as a young geologist: "The prospect of finding oil was so bleak that on arrival I was told that it was unlikely that I would be in Arabia for more than a year," he said. "We all expected to be shifted to Indonesia."[9]

In March 1938 one of the dry oil holes was redrilled, and out gushed great quantities of oil. Quickly, sur-rounding holes were drilled and other fields discovered.

Today Saudi Arabia is the largest oil-producing country in the world and also has the largest known oil reserves.

Money and the things money can buy are seen everywhere. Whole cities have been built in the desert sand. New Mercedes cars wait outside bedouin tents. Well-planned, elaborate college campuses are bulging with students.

What is true in Saudi Arabia is true also in the other countries on the Arabian Peninsula, the land mass between the Red Sea and the Arabian Gulf. The town planner of Abu Dhabi, the capital of the United Arab Emirates, speaks glowingly of future projects: "Our new Sports City, 250 million dirhams [63 million dollars]; our new Summit Conference City, with 50 villas for heads of state, 20 million dirhams; our new 'Wall Street' area, already under construction; a new satellite city with free houses for U.A.E. citizens. There'll be shopping malls, lots of green. If we ever want a university, it should be sited there."[10]

There are other plans: a new jet airport, a new beach road, a second bridge to link the island city with the mainland, a new police-and-defense-force housing complex.

Is there any question that this area is fast becoming the money center of the world? Already the three wealthiest nations in the world—Kuwait, United Arab Emirates, and Qatar—are on this Arabian Peninsula. And the largest country, Saudi Arabia, is not far behind.

How interesting all this appears when compared with the Scriptures.

Although Arabia is mentioned very few times in the Bible, it is often mentioned as a place of much commerce and great wealth.

Until recently, when one thought of Saudi Arabia he thought of one of the most barren countries of the entire world. Although it is presumed part of Asia, Saudi Arabia has a climate and terrain similar to the Sahara of Africa. Some consider it an extension of the Sahara. There are no permanent streams that reach the sea, so lack of water is a major problem. The important agricultural products are dates and grain sorghums.

The Arabia portrayed by Scripture is not a dry and barren Arabia. There are several interesting parallels between the Arabia of Scripture and the Arabia of today. The Arabia of Scripture was a land of great wealth.

Naphtha storage tanks are located at Ras Tanura terminal in Saudi Arabia. Photo courtesy of Aramco.

This was especially true during the reign of Solomon, for the kings of Arabia brought a great deal of gold to him (1 Kings 10:15; 2 Chronicles 9:14). Even the Queen of Sheba, who was from southern Arabia, brought gold, spices, and precious stones to Solomon (1 Kings 10:2-10). Much of Arabia's wealth at that time came from the caravans and commerce that traveled across her land. Some believe the Queen of Sheba went to Solomon to request protection for the vital spice caravans of Sheba.

Arabia's wealth is mentioned in other passages as well. The psalmist speaks of the kings of Sheba offering gifts as well as the "gold of Sheba" (Psalm 72:10, 15). The prophets also mention gold and incense that came from Sheba (Isaiah 60:6; Jeremiah 6:20; Ezekiel 27:22-23). Without question, Arabia was an important source of gold, precious stones, and incense during Bible times.

Such wealth, however, has greatly intensified today. So much money has come into this area that a whole new society is being built. To plan carefully for the future, Saudi Arabia has set goals for developing the country in stages. The major objectives for the kingdom are (1) to increase the growth rate of the gross national product, (2) to develop human resources so all may fully participate in the development process, and (3) to diversify the sources of national income and reduce the country's dependence on oil by increasing the contribution of other production sectors to the gross national product.

To reach these objectives, Saudi Arabia must expand

Ninety-six wheels carry a one-piece, 500-ton pressure vessel to its erection site in Saudi Arabia. Photo courtesy of Aramco.

the facilities in all stages of education. The country must extend health care in prevention, training, hospitals, and health centers. It must build roads and highways, increase the ports, and improve mail delivery. It must increase industry, agriculture, and commerce. Saudi Arabia must change its entire society.

Another similarity between present-day Arabia and the Arabia of Scripture is the emphasis on camels. In the Midianite raids against Israel, "their camels were without number" (Judges 6:5). The people even decorated their camels with necklaces of gold (Judges 8:26). This should not be surprising. One of the first mentions of Arabs in the Bible shows they were camel riders and carried spices and incense on their camels to Egypt (Genesis 37:25).

Throughout history the camel has been necessary to the society of Arabia. The camel has provided transportation, milk, food, and even companionship. It has been a beast of burden and an instrument of war. (Herodotus mentions Arabs fighting from their camels, and Assyrian bas reliefs depict this.) The camel has fit in with the Arabian nomadic and tribal type of life because the camel is an animal of endurance.

The dependence on camels, however, is a great problem in Arabia today. The leaders are trying to move a camel society into the modern world. It is not unusual to see camels and tractors working on similar projects, or a bedouin tent with camels in the back and a new European car in the front.

The problem is intensified by the demands of the most ambitious peacetime construction program ever attempted. On the outskirts of dusty Riyadh, the capital of Saudi Arabia, a new campus for the University of Riyadh is being built. The complex will have more than ten million square feet of teaching space, more than at Harvard. An industrial city for 200,000 people is being built at the tiny fishing village of Jubail on the Arabian Gulf. Moreover, the airport that will be built for the city of Riyadh will be larger than London's Heathrow or Amsterdam's Schiphol. It will be so large that the mosque in the middle will accommodate eight thousand people.

How can an instant city be built? Even in a highly civilized society such tasks would be difficult. What about in Saudi Arabia? What about on the Arabian Peninsula? These countries lack the social substructure that we take for granted. They lack roads, ports, com-

Saudi Arabia is now growing broccoli as an experimental crop. Photo courtesy of Aramco.

41

munications, electricity, housing, and supporting industries. They lack schools to teach literacy, the discipline industry requires. The Arabian society has found it almost impossible to change from camels to other transportation. Oil can be discovered overnight, but changing people takes much longer.

Scripture also shows that Arabians, at various times, have fought against Israel. For instance, they raided Jerusalem during the reign of Jehoram (2 Chronicles 21:16-17; 22:1) and were later defeated by King Uzziah (2 Chronicles 26:7). When Nehemiah was rebuilding the city of Jerusalem, he was hindered by Geshem the Arabian (Nehemiah 2:19). Later, the governor of Damascus, representing the Nabataean king Aretas IV (9 B.C.—A.D. 40), set guards to catch the apostle Paul (2 Corinthians 11:32-33). Nevertheless, the apostle escaped because of the quick thinking of his companions.

In light of this, does it seem unusual for the Arabian countries to be buying arms rather than giving alms? Saudi Arabia is one of the world's largest armament consumers. The weapons they buy are the very latest and most sophisticated. Although the ostensible purpose of having these weapons is to protect the new oil lands, can anyone say they will not be used against Israel eventually?

The Bible also shows another characteristic of the Arabians. This is the most important quality of wisdom. Solomon's wisdom was compared to the wisdom of the "children of the east country" (1 Kings 4:30), the land of Arabia. Moreover, in the book of Proverbs,

LPG — liquefied petroleum gas — forms an important part of Saudi Arabia's hydrocarbon exports. Photo courtesy of Aramco.

which was written mostly by Solomon, there are contributions from two kings of Massa, a tribe of Ishmael (Genesis 25:13-14). The kings were Agur (Proverbs 30:1) and Lemuel (Proverbs 31:1), and both sections are words of wisdom. Many also believe that the wisdom book of Job reflects the background of northwest Arabia.

The Arabian struggle for wisdom continues today. Many of the leaders in the oil industries are graduates of American universities. In Saudi Arabia government schools provide elementary and secondary education for boys. There are also agricultural, industrial, and commerce institutes. Free education is provided in all stages and levels along with textbooks and other school materials. In addition, students in certain stages receive monthly allowances. There is free health care and a free meal program. Scholarships abroad for advanced degrees are also granted, and students have studied in Europe, America and many other countries.

What is the future of Saudi Arabia? The present looks very bright in spite of a few problems. Never before has this country had so much money. With this money, Arabia has formed a new society.

What does the Bible predict about Saudi Arabia? One passage says that one of the Arabian tribes would diminish (Isaiah 21:13-17), but this prophecy was fulfilled in Old Testament times. Another passage mentions that the kings of Arabia would drink the cup of divine judgment (Jeremiah 25:15-19, 24). This also was fulfilled. A third passage predicts that the Arabian would not pitch his tent in Babylon after its destruction (Isaiah 13:20). This, too, has been fulfilled.

The future of many nations is mentioned in Scripture, especially those that surround Israel. And Saudi Arabia is one of the largest and closest land masses near Israel. Why, then, are there no biblical passages that predict the future of Arabia?

One possible reason the future of Saudi Arabia is not mentioned is that she may be destroyed because of her wealth. The vast riches of this nation may be a hindrance rather than a help. The Bible repeatedly warns people to be content with what they have, because riches only bring temptations. The rich young ruler's love of money was his besetting sin, and he turned from Christ (Luke 18:18-25). Judas Iscariot sold his Lord for thirty pieces of silver (Matthew 26:14-15). Although Scripture does not condemn the possession of wealth, it does point out that we are to be stewards rather than owners of riches. We are to dispense what we have for the glory of God and with proper regard for the needs of others. Entire nations, like Tyre, who put their trust in riches instead of the Lord may be destroyed by God (Ezekiel 26).

The reason Saudi Arabia is not mentioned in the future program of God, however, is that this entire area will one day be taken over by Israel. God promised that Israel will have the land between "the river of Egypt unto the great river, the river Euphrates" (Genesis 15:18). When and how this will happen is left to a later chapter.

Are there other Arab nations specifically mentioned in the future program of God? Yes indeed, and we turn to an important one now.

The port of Shuwaikh, Kuwait, is always busy.
Photo courtesy of Arab Information Center.

4

Will Egypt's Glory Return?

Eight thousand miles from New York City stands a building that houses some of the most valuable treasures ever found. Steel bars protect all the windows as black uniformed guards pace back and forth. Large buses arrive hourly bringing thousands of visitors from all over the world.

This is the Cairo Museum in Cairo, Egypt. It contains rich treasures of the past. Probably the most important artifacts are the contents of King Tutankhamun's tomb, which line two galleries. This famous king lived 3½ millennia ago at an important time in Egyptian history. When he died, many of his treasures were buried with him. His tomb was not discovered until 1922.

To see the size and location of the tomb, I visited the Valley of the Kings in Luxor, Egypt, nearly four hundred miles south of the city of Cairo. I walked down the long corridor of the tomb to the burial chamber and antechamber and relived the thrill of Howard Carter when he first uncovered the king's

Hieroglyphics decorate the walls of the Amon-Ra Temple, Luxor, Egypt. Photo by Edgar James.

treasures. I returned to Cairo and inspected the contents of the tomb displayed in the Cairo Museum. There are gilded cows' heads, golden shrines, furniture, bracelets, necklaces, and even a gold mask.

All of these treasures are from the golden age of Egyptian civilization that existed some fourteen hundred years before Christ. It was then that Egypt established her rule as far northward as Syria and as far southward as the Sudan. Tax money from these areas flowed into Egypt, natural resources were developed, gold was plentiful. Egypt had been a wealthy nation, in her glory, a world power.

But since those golden days the Egyptian empire has diminished, the civilization decayed, and the wealth disappeared. Today Egypt is a nation of 38 million people, most of whom live in extreme poverty. The nation faces limited resources and overpopulation. To stabilize the prices of essentials, the government uses 50 percent of its revenue to subsidize basic commodities. Egypt is dependent on the handouts from other nations just to keep her head above water.

Recent events, however, may signal a change in the fortunes of this land of the Nile. The Suez Canal is again open and contributing millions in revenue each year. Oil production is increasing, and new discoveries are being made. Says an American oil engineer: "I . . . guess that in 10 years Egypt will be a producer on the scale of Kuwait."[11] Most important is this nation's rise as the leader of the Arab world.

This is interesting in light of the predictions in Scripture. Years ago the prophets mentioned such a turn of events for this troubled land. The prophet Isaiah said that "the LORD shall smite Egypt: he shall smite and heal it" (Isaiah 19:22). He spoke not only of Egypt's judgment, but also of her blessing. One day the land will be healed.

The recent incidents surrounding Egypt's latest rise to power raise important questions about her future. Are these rumblings the beginning of a new Egyptian empire? Is this the foreshadowing of the "healing" of Egypt? Does the Bible predict that Egypt's glory will return?

Outside of Israel, the most important country mentioned in Scripture is the nation of Egypt. It is first mentioned as Mizraim (Genesis 10:6), who was one of the sons of Ham. The word *Mizraim* is dual, and many believe it refers to the natural division of the country into an upper and lower region. Some Egyptians, however, have referred to their land by the name "Kemet," which means "the black land." This rich land, irrigated by the Nile River, has been the source of the wealth of the nation down through history.

The land, the "gift of the Nile," was fertilized by deposits from the river. The land was protected by the sea in the north, deserts and mountains on each side, and valleys and cataracts on the south. Because of the climate, it is possible to grow three crops a year. Agriculture is the foundation of Egyptian economy.

The civilization of Egypt was profound by any standard. The people had ways to make cloth, vessels, and pyramids of such delicacy and magnitude as has never been equaled. The greatest of the pyramids at Giza was built by Khufu about 2500 B.C. This pyramid was originally 492 feet high, 755 feet square at the base, and covered almost thirteen acres. It contained 2,300,000 blocks of limestone, each one weighing about two and a half tons.

The religion of Egypt was a polytheistic worship of the gods of nature. Re, the sun god; Osiris, god of the Nile; and the moon were all worshiped. Egypt had a conglomeration of many deities, and at one time each town within Egypt had its own god. The Egyptians believed that every object was indwelt by a spirit that could choose its own form, thereby occupying the body of a crocodile, a fish, a cow, or another animal. The Egyptians had numerous holy animals and faithfully regarded them as such. There was, however, a remarkable understanding of morality and mortality. Unusual steps were taken to insure the welfare of the deceased in the life hereafter. This is why the kings protected themselves with such elaborate tombs.

What predictions does the Bible give concerning this nation? One of the most important prophecies has already been fulfilled. It pertained to the judgment of Egypt when Israel was delivered from slavery.

Egypt became a refuge when Abraham went down to Egypt to escape famine in Palestine (Genesis 12:10). Crop failure was rare in Egypt because of the unusual climate. Even during Roman times wheat was

In Egypt, where the Nile River is, so is civilization. A person can put one foot in the Sahara Desert and the other on green grass. Photo by Edgar James.

exported to Italy, so it was natural for Egypt to be the breadbasket for surrounding nations.

When Abraham went down to Egypt, he lied about his wife, Sarah. He told her to say she was his sister. Going to Egypt provided an occasion for sin. Another result of this visit was that Hagar, Sarah's maid, was brought back from Egypt, and she became the mother of Ishmael. His descendants have ever since plagued the Israelites.

Famine brought Jacob and his sons to the land of Egypt. One of the sons, Joseph, had already been sold by his brothers to a group of Ishmaelites and had been brought to Egypt. While there, he was highly exalted because of God's wisdom. When his brothers came, he revealed himself to them, and they rejoiced greatly. God had predicted that not only Abraham, but all Israel would go down to Egypt and "serve them; and they [Egypt] shall afflict them [Israel] four hundred years" (Genesis 15:13). God also predicted that He would judge Egypt for enslaving His people. "And also that nation, whom they shall serve, will I judge: and afterward shall they come out with great substance" (Genesis 15:14).

The deliverance of God's people from the land of Egypt is a major prophetic theme of the Old Testament. There are over 125 references to the deliverence. At the time the prophecy was given, rescue did not seem likely. But in the time of need, God raised up Moses and Joshua to deliver His people, and the army of Egypt perished in the sea (Exodus 14:27).

51

Egypt continued to be an important power after Israel left. During the time of Solomon the land of Egypt was a place of refuge for the king's political enemies. On one occasion, Hadad, an Edomite, fled to Egypt and became active against Solomon (1 Kings 11:14-20). On another occasion Jeroboam went to Egypt to escape Solomon's wrath. Solomon, however, did trade with Egypt (2 Chronicles 1:16-17), and his wisdom surpassed even that of Egypt (1 Kings 4:30).

Another important prophecy already fulfilled concerns another judgment of Egypt. Nine hundred years after Israel was delivered from Egyptian bondage Egypt was still a mighty nation. But in addition, there was Babylon in the north, vying for world leadership. Israel was between these two countries, and she was attacked and oppressed by both of them. God promised judgment on these nations.

God used Nebuchadnezzar, king of Babylon, to judge Egypt. In fact, Jeremiah predicted that "Nebuchadnezzar king of Babylon should come and smite the land of Egypt" (Jeremiah 46:13). Ezekiel echoed this fact. He said that God would "give the land of Egypt unto Nebuchadnezzar king of Babylon; and he shall take her multitude, and take her spoil, and take her prey" (Ezekiel 29:19).

For many years no one could find any evidence that Babylon had ever taken Egypt. Finally, in the latter part of the nineteenth century, indisputable evidence was found on a Babylonian tablet now housed in the British Museum. The tablet reads, "In the thirtyeventh year, Nebuchadnezzar, king of Babylon marched against Egypt to deliver a battle." Later, the Medo-Persian Empire completed the destruction begun by Nebuchadnezzar. Still later Greece seized the glory that once had been Egypt's.

Such a devastating destruction was also predicted by Scripture. Egypt was to be "the basest of the kingdoms; neither shall it exalt itself any more above the nations: for I will diminish them, that they shall no more rule over the nations" (Ezekiel 29:15). Moreover, the Egyptians were to be scattered "among the nations, and disperse them among the countries; and they shall know that I am the LORD" (Ezekiel 30:26).

This is why Egypt has been such a decadent nation through the centuries. There have been a few times, as during the Arab rule, that the country was stirred. But Egypt has never regained her past glory.

Has God finished judging the nation of Egypt? Not according to Scripture. Although God judged this nation when Israel was in bondage and again when Israel was oppressed, He will judge this nation again. Even though this nation is beginning to raise its head today, God says that "the sword shall come upon Egypt," and there will be a time when "the slain shall fall in Egypt, and they shall take away her multitude, and her foundations shall be broken down" (Ezekiel 30:4). All of this will happen during the Day of the LORD (Ezekiel 30:3). This will immediately precede the second coming of Christ, when there will be great devastation

The Nile river runs through the modern city of Cairo, Egypt. Photo by Edgar James.

upon the earth. Egypt will be part of this time of trouble. The idols will be destroyed, and God will put a fear in the land (Ezekiel 30:13).

The Antichrist will come against Egypt when he enters the land of Palestine and fights against Israel. This world dictator will take the spoil of the country, and even "the land of Egypt shall not escape" (Daniel 11:42).

Why is there so much judgment against Egypt? Why would God destroy a nation that contributed so much to the world's civilization?

One possible reason is that Egypt is close to Israel and therefore should have known the ways of the Lord. Egypt saw God work in Israel, and Egypt should

54

ave also obeyed Him. With knowledge always comes
sponsibility. This nation, which has seen the tes-
mony of the Lord, will be again brought down.

Another reason may be her oppression of Israel.
ong before Israel was enslaved, God had said, "I will
less them that bless thee, and curse him that curseth
hee" (Genesis 12:3). This has been true through the
enturies, and it has been true for the nation Egypt.
gypt has been judged for her oppression of the
ewish people, and she may yet be judged in the fu-
ure.

But perhaps the greatest reason is Egypt's sin of
ride. This nation depends on the Nile River and yet
loes not give God glory. She says, "My river is mine
wn, and I have made it for myself" (Ezekiel 29:3).
She, like other nations, looks to herself and does not
glorify the Lord, who has had His hand upon her. This
s why God has said, "And the land of Egypt shall be
lesolate and waste; and they shall know that I am the
LORD: because he hath said, The river is mine, and I
have made it" (Ezekiel 29:9). If this is true for Egypt,
will it not also be true for the others?

God is a God of justice, and He must punish sin. But
He is also a God of grace. Is the picture therefore
totally black? Is there not a ray of hope? Is this nation
to be removed from the face of the earth? Not according to Scripture. The Bible predicts not only judgment
for this nation, but also vast blessing. One day God will
restore the land of Egypt, and all in it will know the
Lord. The prophet Isaiah said, "The LORD shall smite

Egypt: he shall smite and heal it: and they shall return even to the LORD, and he shall be intreated of them and shall heal them" (Isaiah 19:22). In other words there will not only be judgment but also blessing.

Could what is happening in Egypt today be the beginning of that healing? The Suez Canal is fast regaining its status as one of the world's most important waterways. Egypt's oil prospects are quite good. Many foreign firms are searching for oil in the Sinai, the Red Sea, the Gulf of Suez, and the Western Desert. Tourism is rapidly increasing, and new hotels are accommodating the rising tide.

Egypt's future looks brighter today than it has for many years. But the Scriptures point out that when God heals this land, accompanying that healing will be a great outpouring of the knowledge of the Lord. The prophet writes, "And the LORD shall be known to Egypt, and the Egyptians shall know the LORD in that day" (Isaiah 19:21). It is evident that this is not true today, nor has it ever been true in the history of Egypt. This is unfulfilled prophecy, although the time may be close for it to be fulfilled. What is happening today should show us that God is not through with Egypt or the Middle East. Egypt, after being dead for so many years, is coming back to life. Not only is a future judgment coming, but also a future healing. The Lord will keep His Word.

What will happen when God heals Egypt? Egypt and the Gentiles will know the Lord's rest (Isaiah 11:10). Isaiah speaks of a day when righteousness will rule, and the wolf will dwell with the lamb. This will be when Christ will rule this earth, the time after He returns. It is "in that day" that there will be a signal for the people, and they will know the rest of the Lord. When Christ, the Prince of Peace, reigns upon the earth, the world will experience true peace and rest at last.

Another event that will occur when God heals Egypt is that Jews will be regathered from Egypt (Isaiah 11:11). This is also said to be "in that day," and therefore must also be at the time Christ returns. What is interesting is that God will "the second time" recover His remnant. When Israel was taken captive into Babylon, God recovered His people and brought them back into the land. Israel was again taken into captivity by Titus and his Roman armies in A.D. 70, and the city of Jerusalem was destroyed. Although some Israelis have come back to the land today, God will regather them when He comes again. This regathering will be from many nations, including the land of Egypt.

It is in that day that "the tongue of the Egyptian sea" will be destroyed (Isaiah 11:15). To what does that refer? There are three bodies of water of major significance that are associated with the nation Egypt—the Nile River, the Mediterranean Sea, and the Red Sea. A large body of water like the Nile River could be spoken of as a sea, but most Bible scholars hold that that is not what is meant here. The Mediterranean Sea is never referred to as "the Egyptian Sea," either in Scripture or in other books. "The tongue of the Egyptian sea" must refer to the Red Sea, specifically the northern part that runs into the Mediterranean. This is what would be the "tongue" of it, the part that runs through the Suez Canal. Although this canal is one of the world's most important waterways, this area will one day be destroyed by the Lord. Perhaps this will be in conjunction with the highway that will be built through Israel to the land of Assyria (Isaiah 11:16; 19:23).

There are other events associated with God's healing of Egypt. Some of these, as we have seen, come from the great millennial passage, Isaiah 11. It is during the Millennium that Christ will be reigning, it is then that the curse on the earth will be lifted, it is then that God will again deal with Egypt and other nations.

The other events are predicted in Isaiah 19, a passage that talks about the nation of Egypt. Isaiah 19 begins with the words, "The burden of Egypt." Although this passage speaks of the past, it also speaks of the future. God says He will heal Egypt (v. 22) and describes events that will accompany that healing. Six times when speaking of the healing of Egypt, God uses the phrase "in that day," meaning at the time of Egypt's healing (vv. 16, 18, 19, 21, 23, 24).

When Egypt is healed, the people will be afraid because of the shaking of the Lord of hosts (v. 16). This must refer to the Lord's future judgment of Egypt. Before there will be true healing, there will be the judgment of the Lord.

Perhaps God is going to use Israel in His judgment

of Egypt, for "the land of Judah shall be a terror unto Egypt" (v. 17). Moreover, the language of Israel will be used in Egypt, for "five cities in the land of Egypt speak the language of Canaan" (v. 18). Today there is deep hatred between Israel and Egypt. It is impossible to travel directly between these two countries, let alone speak Hebrew in Egypt. But judgment comes before blessing, and the Lord will in some way use the nation of Israel in His judgment of the people of Egypt.

Yet there is hope, there is promise, there is deliverance. God will send unto them "a savior, and a great one, and he shall deliver them" (v. 20). This is why "the Lord shall be known to Egypt, and the Egyptians shall know the Lord" (v. 21). This is not true today. Egypt is a Muslim country. But one day that will all be changed. When Christ returns, those in this country will know the Lord. This is why God will be able to heal this country.

One evidence of that knowledge of the Lord will be an altar to the Lord "in the midst of the land of Egypt" (v. 19). Although Egypt has been known for her pyramids and tombs that have recorded past civilizations, in the future she will be known for her altar to the Lord. Sacrifices may also be reinstituted (v. 21). If so, this altar may be used for that purpose. Of course, such sacrifices would not have the same purpose that sacrifices had during Old Testament times, when they covered sin until Christ died. Rather, they would be a memorial of Christ's death, and this testimony would be spoken of throughout the land.

Not only will this be a time of judgment and deliverance, but also a time of great blessing. As Israel served the Egyptians, now the Egyptians will serve the Lord (v. 23). Moreover, Israel will be "the third with Egypt and with Assyria" (v. 24). This does not mean that Israel will be third in order of priority, but that there will be three nations that will especially enjoy the blessing of God—Israel, Egypt, and Assyria.

One tangible evidence of this blessing will be the highway built between Egypt and Assyria (v. 23). A careful study of a map of the Middle East shows that this highway must go straight across the land of Palestine. This is evidence that there will be peace. Perhaps the tongue of the Egyptian sea will be destroyed so this highway can be constructed.

Although Egypt will be at peace with Israel, there will need to be continued discipline during the Millennium. During the Millenium the Feast of Tabernacles will be held, and God promises punishment to Egypt if she does not keep the feast (Zechariah 14:18-19).

The purpose of God, then, is one day to bless Egypt along with Israel. This nation, full of slavery and sin, will one day witness the grace of God. It is then that the Lord Himself will say, "Blessed be Egypt my people" (Isaiah 19:25).

5

Black Gold in the Northern Hills

On a rocky plain in northern Iraq there burns an eternal fire. Its flames cover an area ten yards wide. The fire burns like a grass fire, but it always stays in one place and is much, much hotter.

No one knows when this fire began. Some think it predates all history.

It may have been used to "burn thoroughly" the bricks made nearby that were used to build the tower of Babel. Or it may have been part of the fiery furnace into which Daniel's companions were thrown. But whenever the fire began or for whatever reason, there is no question it is due to a rich seepage of natural gas.

Oil companies began drilling near this eternal fire. As a result, on October 14, 1927, two plumes of oil shot 140 feet into the air and flooded the countryside. The first well was called "Baba Gurbur," meaning "eternal fire," since the well was located two miles from the fire.

"It proved to be one of the most remarkable strikes ever made,"[12] a veteran oil man said in his Baghdad office. It took ten days to bring the gusher under control. Men worked around the clock to cap the well that would produce 80,000 barrels a day. They feared a spark might ignite the gas and blow everything up. Meanwhile, oil accumulated all over the countryside.

All of this happened in Iraq, a nation north of modern Israel. The country of Iraq is backward by today's standards. Although it was once the "cradle of civilization," it is now a troubled land. It seems to fulfill the old Arab proverb which says, "Yemen is the cradle of the Arab race and Iraq is its grave."

Many different nationalities have tried to survive here. Amid the marshes of the Tigris and Euphrates Rivers in the south live the Maadans, a marsh-dwelling people descended from the Sumerians, Babylonians, Persians, and Arabs. In the north are the Kurds, people who trace their origin to the ancient Medes.

Religions also vary here. Most people are Muslim, but some are Nestorian, Chaldean, or Assyrian Christian. The Muslims are divided into Sunni and Shiia groups. The Sunni are mainly urban dwellers and in positions of leadership; the Shiia are farmers and village dwellers. Poorly educated, they are easily exploited.

"Iraq is not a country in the Western sense of the word," a diplomat said. "It is a hodgepodge of nationalities and religions which have been tossed together by fate and history."[13] It can be ruled only through the strong central government in Baghdad, one ruthless in seeking its ends.

Is this bleak assessment changing? With the discovery of oil, much has happened in Iraq. It is now open to Western influences. It has changed from a backward country torn by tribal feuds to one with a sense of nationalism. It has become an outspoken country in the Arab's fight against Israel. Iraq has been aligned with Russia, and has been given a destiny. Oil has given it an affinity with Syria and Lebanon since oil pipelines from Iraq travel through those countries to the Mediterranean.

All this is interesting when compared to the Scriptures. God said that prophecies of judgment and destruction will one day be replaced by prophecies of blessing. One day the "Egyptians shall serve with the Assyrians" (Isaiah 19:23). Present-day Iraq was ancient Assyria. One day God will call this people "the work of my hands" (Isaiah 19:25). Could what is happening today be the beginning of God's blessing on Assyria?

Iraq was once the center of ancient history. Some believe the Garden of Eden flourished there. The wheel was invented there. The earliest writing was done there. The earliest known code of laws was written there. The earliest known university was built there. Irrigation, money, and the arch employed in building, were all used there first.

Iraq has been the center of important past civilizations. Some of these were the Sumerian, Akkadian,

A new Middle East pipeline is wrapped with pro-
tective coating before it is lowered into its trench.
Photo courtesy of Aramco.

Amorite, Babylonian, and Assyrian nations. The
Sumerians formed city-states there three thousand
years before Christ. They had a keen sense of private
property and were ardent enterprisers in trade. They
believed the gods owned the land, and they worked to
please the gods. They made discoveries in irrigation,
law codes, and the use of money, all of which are used
today.

The Babylonians lived in the southern part of
present-day Iraq, in the area between the Tigris and
Euphrates Rivers. They built the city of Babylon, which
was their capital, and developed an empire that became the terror of western Asia. This was sometimes
called "the empire of Nebuchadnezzar."

The Assyrian Empire was located in the northern part of present-day Iraq. This nation is first mentioned in Scripture when the Garden of Eden is described as being near the Tigris, or Hiddekel River. It is said that this river goes "toward the east of Assyria" (Genesis 2:14).

Assyria was founded by Babylonian colonists. The early capital, Asshur, was dedicated to the national god, who was the incarnation of war. One of the major cities of Assyria was Nineveh, where the prophet Jonah was commanded to go.

What prophecies does Scripture give concerning these past peoples? Many of these nations, like the Sumerians and Akkadians, were important, and archaeologists have discovered their contributions to civilization, but the Scriptures are silent regarding them.

The Bible does give many prophecies concerning two of these empires, however. One of the earliest prophecies concerned Babylon, which was founded by the descendants of Cush and followers of Nimrod (Genesis 10:8-10). The people of Babylon decided to build a tower, a man-made bridge to God. They used bricks and mortar, but God restrained them from making this false approach to heaven. He said He would go down and "there confound their language, that they may not understand one another's speech" (Genesis 11:7). The Lord not only confounded their language, but scattered them upon the face of the earth.

Years later, another prophecy regarding Babylon was given. This prophecy told how Babylon would forcibly take the Jews into captivity. This is the subject of several of the prophets, including Isaiah, Jeremiah, Ezekiel, and Daniel.

Isaiah predicted the Babylonian captivity when Babylon was a rather obscure power and gave no indication of its future greatness (Isaiah 13-14, 47). In contrast, Jeremiah's prophecies were given when Babylon was a great power and had no indication of its coming destruction. Yet, Jeremiah said Babylon would be destroyed because of the way she treated Israel (Jeremiah 50, 51). The prophet also predicted that the captivity would be for seventy years (Jeremiah 25:11-12).

Ezekiel confirmed these predictions and added that the city of Tyre would also be destroyed (Ezekiel 26-28). Daniel spoke of the Babylonian leader God would use to bring about the destruction of the nations. The leader was Nebuchadnezzar (Daniel 2, 7). The completion of all these prophecies shows how God meticulously fulfills His Word.

The Bible also predicts the future of the Assyrian Empire, the empire that once was located in the north of present-day Iraq. Many rulers of Assyria conquered surrounding cities and thereby put together this great empire. Some of these rulers were Tiglath-Pileser I (1114-1076 B.C.), Shalmaneser III (858-824 B.C.), and Tiglath-Pileser III (745-726 B.C.). It was during this time that the empire went all the way to the Mediterranean Sea, and Syria and northern Israel were annexed to it.

Isaiah the prophet predicted that Assyria would in-

The flow of crude oil is controlled from this room.
Photo courtesy of Aramco.

vade Israel and take the people captive (Isaiah 7, 8). He
also predicted that Assyria, because of God's judgment,
would be punished and brought down (Isaiah 11:12-
16).

Sargon II (722-705 B.C.) of Assyria came against
Israel and deported twenty-seven thousand Israelites to
cities in Assyria. He then replaced those people with
natives from Syria and Babylonia. The next ruler,
Sennacherib (705-681 B.C.), took the city of Babylon
and deported more than two hundred eight thousand
people from there as captives. He traveled to the
Mediterranean coast, took tribute from Phoenicia, and
captured forty-six cities of Judah. He is credited with
having deported over two hundred thousand people,
and he shut up Hezekiah "like a bird in a cage" in
Jerusalem.

But God kept His Word. The great Assyrian Empire gave way to the Neo-Babylonian Empire, and the city of Nineveh fell to the Babylonians in 612 B.C. During its day, however, the Assyrian Empire was a vast empire that extended all across the northern part of Israel.

Other important prophecies about the Assyrian Empire have been fulfilled. These are contained in the books of Nahum and Jonah. The entire book of Nahum speaks of the destruction of Assyria, especially the city of Nineveh. The book of Jonah discusses the remarkable experience of the prophet Jonah in going to warn the city of Nineveh to repent of its sins. Because of their repentance, the city was spared for 150 years.

As God worked in the south of Israel with Egypt, so God worked in the north of Israel with Babylon and Assyria. The prophets foretold what God would do, and God kept His Word.

What about today? Are there any prophecies regarding these nations in the north that will be fulfilled? And if so, what are they?

Many believe a complete destruction of Babylon will take place at the second coming of Christ. Although Babylon was destroyed when the Medes and Persians took it over, the city continued to flourish in one form or another until A.D. 1000. Yet the prophet said, "It shall never be inhabited, neither shall it be dwelt in from generation to generation: neither shall the Arabian pitch tent there; neither shall the shepherds make their fold there" (Isaiah 13:20).

It is true that, in a political and religious sense, this city will be destroyed at the second coming of Christ, yet some hold that the only way it could be destroyed physically is if it is rebuilt. Those who believe a future physical destruction will take place find John describing the merchants of the city weeping over her fabulous wealth when she is finally destroyed (Revelation 18:9-19). Could it be that the present resurgence of oil-rich Iraq means the reconstruction of the ancient city of Babylon?

The importance of Assyria is seen by the fact that there are over one hundred forty references in the Bible to this people, with over twenty references to the principal city, Nineveh. Many of these prophecies have already been fulfilled and concern the judgment of Assyria. But God is a God of grace as well as of justice, and a careful study of Scripture shows that God has a future for this important nation.

For instance, one future prophecy of this nation concerns the regathering of Israel from many nations, including Assyria (Isaiah 11:11; 27:13; Zechariah 10:10-11). Although some Jews have already gone back to Palestine, there will come a time when all Jews will be regathered and brought back to Palestine. All Jews will return when Christ returns.

Another prophecy concerning Assyria is that a highway will be built from Assyria to Egypt (Isaiah 11:16; 19:23). This highway will have to cut across the nation Israel, for this is the only way such a highway could go.

This means there will one day be peace among these nations, because the Lord will be reigning on the earth.

The highway between Assyria and Egypt will be used for commerce, travel, and communication. There will be a free flow of traffic between these nations—a sharp contrast to the present scene.

Another prophecy for Assyria shows God's great blessing upon it. Assyrians will serve with the Egyptians in the coming Kingdom. Although some Arab nations will not have a future, according to the Word of God, the Assyrians will. They will be blessed by God, and they will be a "blessing in the midst of the land" (Isaiah 19:24).

Although God especially chose Israel as a people for Himself, He does have a purpose for other nations. He

The tanker *Globtik Tokyo* takes on a cargo of crude oil. Photo courtesy of Aramco.

points out that Israel will be "the third with Egypt and with Assyria" (Isaiah 19:24). This does not mean that Israel will be third in order of priority but that there will be three important nations in the coming Kingdom.

The Assyrian's hope is in God, not in oil. One day the Lord will say, "Blessed be Egypt my people, and Assyria the work of my hands, and Israel mine inheritance" (Isaiah 19:25). Assyria does have a future under the hand of God.

6

Land — the Goal of the Palestinians

In an Arab state almost in sight of Israel, twenty-four commandoes undergo intensive military training. They wear camouflaged, khaki uniforms as they learn techniques of sabotage and guerrilla warfare. They listen intently as the instructor teaches the art of plastic explosives. They automatically drop flat on the ground as another instructor coolly lays a barrage of machine-gun fire over their heads.

These are Palestine liberation trainees. Barely eighteen years of age, they learn sabotage, terrorism, and guerrilla warfare as a way of life.

"We are fighting for the return of our land," insists Colonel Abdulah Wajih. He is a graduate of Syria's military academy and assigned to training PLO-affiliated *fedayeen,* or guerrilla, groups. "We could live in peace with the Jews if our land were returned to us," he says. "Over there is Palestine," and that is what he wants. He adds, "I am a Palestinian. My six children are

Palestinian. We will be Palestinians as long as we live."[14]

What is a Palestinian? The Israelis call them Arabs, but not Palestinians. When the Jews were under fire, they stood and defended their homes. But the Arabs ran, former Prime Minister Golda Meir pointed out. "We did not chase them out. They left of their own free will. Again I say to you, there are no Palestinians."[15]

Some Israeli leaders might not want to use the word *Palestinian*, because the use of the term might suggest that these Arabs have a claim on the land of Israel. But no matter what one calls them, these Arabs are still refugees, and they pose one of the biggest barriers to the settlement of the Arab-Israeli problem.

Most of these refugees have settled in the west bank, the Gaza strip, and the Hashimite Kingdom of Jordan. But their goal is the land of Israel—they want the land!

Whose is this most contested piece of real estate? The Arabs are struggling for it, the Palestinians claim it is theirs, but the Israelis believe they have the title deed to it. Their basis is a Bible passage written years before the present conflict.

God promised Abraham a covenant if he would leave his country, kindred, and father's house. God helped Abraham along, he obeyed, and God did make a covenant with him. In the Abrahamic covenant, God promised blessing to Abraham, the families of the earth, and the nation Israel. Notice what God promised to Abraham: "I will make of thee a great nation, and I

73

The priceless Dead Sea Scrolls were found in the Qumran Caves on the shore of the Dead Sea. Photo courtesy of the Israel Ministry of Tourism.

will bless thee, and make thy name great; and thou shalt be a blessing" (Genesis 12:2).

God fulfilled this promise literally in every way. From Abraham came a great nation, the nation Israel. Although nations rise and fall, the Jews have continued through history.

God blessed Abraham, not only with a material blessing as he had had in Ur of the Chaldees, but also a spiritual blessing. Abraham believed in the Lord, and it was reckoned to him for righteousness (Genesis 15:6). God also made his name great. Muslims, Jews, and Christians all honor the name of Abraham. Abraham's example of faith, shown throughout Scripture, has been a blessing to many.

God also kept His promise to the families of the

The famous Dead Sea Scrolls are now kept in the Shrine of the Book Museum in Jerusalem. Photo courtesy of the Israel Ministry of Tourism.

earth. He said, "I will bless them that bless thee, and curse him that curseth thee: and in thee shall all the families of the earth be blessed" (Genesis 12:3). Here is the principle that those who bless Israel will be blessed by God, and those who curse Israel will be cursed. This has been true throughout history; one of the most recent examples is the German persecution of the Jews during World War II. Many believe that this may have been one reason Germany lost that war. The last part of the promise, that from Abraham would come blessing, indicates that from Abraham would one day come the Messiah, who indeed has blessed all the families of the earth.

How does this covenant relate to the Palestinian question? God promised that Israel would be a great nation (Genesis 12:2), and that she would also have the land. He said, "Unto thy seed will I give this land" (Genesis 12:7). The first chapters of Genesis show that Israel will one day have the land. In fact, God promised the land as an "everlasting possession" to Abraham and his seed (Genesis 17:8). Abraham asked, "Whereby shall I know that I shall inherit it?" (Genesis 15:8). God confirmed the promise with a sign.

In those days there were many ways to confirm a covenant. After a covenant was made, the parties involved would exchange shoes. This signified that as long as each kept the other's shoes, he would keep the covenant he had made. Another way in which a covenant was confirmed was with salt. The bedouins carried the scarce commodity of salt on their belts. When they made a solemn promise, they would confirm it by taking a pinch of salt and putting it in the other man's pouch, while he did the same. They would then shake their pouches signifying, "Until I give you back your salt and you give me back my salt, we will keep the covenant we have made."

Perhaps the most solemn way to confirm a covenant was with the carcass sign. The two parties would take several animals, kill them, and divide their carcasses into two large piles, with the innards placed on top. Then the two people would link arms and walk between the two piles in a figure-eight pattern. This signified that until the carcasses come back together, they would keep the covenant they had made.

This carcass sign was the way God confirmed His

promise to Abraham (Genesis 15:7-17). God commanded Abraham to make the two large piles of animals. But instead of God and Abraham linking arms, God put Abraham to sleep (v. 12). It was God alone who went between the carcasses (v. 17). This showed that the fulfillment of the covenant depends on God alone, and one day He will fulfill it.

How much land has God promised to give Israel? Only once in Scripture does He give the boundaries, and they are very clear. He said, "Unto thy seed have I given this land, from the river of Egypt unto the great river, the river Euphrates" (Genesis 15:18).

A study of a map shows that these are natural boundaries. The Euphrates cuts across the northern part of the land, almost from the Mediterranean Sea to

Israel's largest city is Tel Aviv. Photo courtesy of Israel Ministry of Tourism.

the Arabian Gulf. The western boundary is the river of Egypt. Some believe this is a reference to a stream near Gaza called Wady-el-Arish, but the Hebrew word in Genesis 15:18 is quite clear. It means "river," and the only real river of Egypt is the Nile.

These boundaries, therefore, include part of Lebanon, Syria, Iraq, and Egypt, and all of Jordan, Saudi Arabia, and the oil-rich Gulf states. God has faithfully kept His promises to Abraham and the nations of the earth, and He will keep His promise of the land to Israel.

Should Israel possess all that land today? The Arabs think that this is Israel's goal and want to hold her back. It is obvious that she does not have all the land. What has happened since the time God promised the land to Israel, and what will happen before Israel will have the land?

The Lord promised not only that Israel would one day have the land, but also that she would be taken from that land. Three times Scripture predicts that Israel would be dispersed from the land and brought back each time. Israel and the land go together. When Israel is out of the land, there is cursing. When Israel is back in the land, there is blessing.

The first dispersion from the land was into Egypt. God said that Israel would "be a stranger in a land that is not their's, and shall serve them; and they shall afflict them four hundred years" (Genesis 15:13). This is exactly what happened when Jacob and his sons went down to Egypt because of famine; their descendants stayed there for four hundred years and were enslaved

78

by the Egyptians. But God also predicted that the Israelites would return to their land. He said, "That nation, whom they shall serve, will I judge: and afterward shall they come out with great substance" (Genesis 15:14). God raised up Moses to lead the people out of Egypt, and Joshua to bring the people back to the land. God fulfilled His Word.

The second dispersion was into another country, Babylon. God said of the divided kingdom, Israel and Judah, that "these nations shall serve the king of Babylon seventy years" (Jeremiah 25:11). God punished Israel by sending her to the land of Babylon, and she was there seventy years. When the time was over, God again raised up leaders to bring His people back. This time He raised up Ezra, Nehemiah, and Zerubbabel, and they brought Israel back into the land.

The third dispersion was also predicted. Instead of the Lord taking Israel into one other nation, however, He said His people would be scattered among all nations. The Israelites were told, "The Lord shall scatter thee among all people, from the one end of the earth even unto the other" (Deuteronomy 28:64).

This dispersion did not occur until A.D. 70, when the Roman general Titus and his armies came against Israel. He besieged the land, leveled the city of Jerusalem, and scattered the Jews to the ends of the earth. Today there are more Jews in the United States than there are in Palestine. There are Jews in England, Europe, Asia, Australia. In fact, the Jews were scattered literally to the ends of the earth.

Women play an important part in the society of Israel. Photo courtesy of Israel Ministry of Tourism.

When will this third dispersion end? Just as God predicted that the Jews would be brought back from Egypt and Babylon, so He has predicted that they will be brought back from the ends of the earth. This will happen when they "obey His voice according to all that I command thee... that then the LORD thy God will turn thy captivity, and have compassion upon thee, and *will return and gather thee from all the nations,* whither the Lord thy God hath scattered thee" (Deuteronomy 30:2-3 italics added).

Since A.D. 70, many things have happened to this land. The Arabs ruled the land from A.D. 637 until the sixteenth century, except during the time of the Crusades in the twelfth and thirteenth centuries. From the sixteenth century until World War I, the Turks ruled the land. In spite of this, however, there have always been some Jews, as well as other immigrants, in the area.

In the late nineteenth century Zionism changed the picture. Zionists claimed that all Jews had a historic right to their homeland of Canaan. Theodor Herzl, longtime Zionist leader, defined its aims at the Basel Congress of 1897: "Zionism strives to create for the Jewish people a home in Palestine secured by public law."

This view was reinforced at the end of World War I by the Balfour Declaration. This British document allowed Jews to return to the land of Palestine. It read, "His Majesty's Government view with favour the establishment in Palestine of a national home for the Jewish people, and will use their best endeavours to facilitate the achievement of this object."

So the way was made, and the Jews came. At the beginning Jewish immigration was small, but it soon became a steady flow through most of the first half of the twentieth century.

Does this mean that the Jews are returning from the third dispersion? Does this mean that they are to take the land?

A careful examination of Deuteronomy 30:2-3 shows that when Israel obeys God's voice, and the Lord returns, the Jews will be gathered from all nations and will possess the land. What is happening today is a foreshadowing of what is yet to take place. But it will not be until Israel has a changed heart and the Lord returns that Israel will be completely restored to the land.

Even those who are against the Zionist movement recognize that Israel will one day have the land. This view is held by Gottfried Neuberger, a representative of an anti-Zionist religious movement. As part of an Arab Symposium he said, "The exile of the Jewish people is divinely decreed, and that they are not permitted to acquire the Holy Land before the appearance of the true Messiah."[16]

What is the present situation? Without question, the Arabs in Jordan, the west bank, and the Gaza strip want land. Yasir Arafat, the militant PLO leader, wants a secular democratic state occupied by both Jews and Arabs. Others, including Elie Eliachar, former Deputy Mayor of Jerusalem, have suggested the creation of a Palestinian state on the west bank and in the Gaza strip.

Eventually, even the city of Jerusalem could be affected. Ben-Gurion, the famed Israeli leader, once predicted, "We'll have to come to terms, even about Jerusalem, for peace."[17]

Within the nation Israel there are some who advocate that instead of giving up land, Israel should be taking more land. One of these is Ariel Sharon, who wants to expand the state of Israel. Sharon says, "One of our goals should be to populate the strip of land from the Golan Heights to Sharm el Sheikh, and I don't mean to put up just twenty or 30 kibbutzim. We

Israel has newspapers and magazines from all over
the world. Photo courtesy of Israel Ministry of
Tourism.

should bring 2 million Jews to this part of Israel by the
year 2000."[18]

The struggle in Palestine today is the struggle for the
land. Both Arabs and Israelis are trying to get all they
can. But God is showing that they cannot do it themselves. This is why outside powers are needed to
guarantee peace. One day God will fulfill his covenant
to Israel just as He fulfilled His promise to Abraham.

One day the Lord will return and give to Israel all the land from "the river of Egypt unto the great river, the river Euphrates." This is why the Scriptures do not predict a future for the nation Jordan. This is why Saudi Arabia and the Gulf states are not mentioned in future prophecy.

Meanwhile, Israel faces great opposition and testing. The Arab oil pressure has political objectives. This is evidenced by the main U.N. General Assembly resolution concerning Israel. This November 1975 resolution declares, "Zionism is a form of racism and racial discrimination." Not long ago Carl Rowan commented that seventy-two countries were openly hostile to Israel, with thirty-two others tottering on the fence.

Such events raise crucial questions. Is the Israel we know today the Israel that will be ringed by foes at the end of the age? Are we beginning to see the pressures that will issue in the great prophetic events to come; the time of Jacob's trouble and the divine deliverance of God's people?

We can only wait and see. One thing is clear: the eyes of the world are turning toward a strategic portion of our globe, the point where Africa, Europe, and Asia intersect, and the place the Bible tells us will be the focus of end-time events.

7

Are Arabs Pawns in Russian Hands?

Syria is at a low ebb. She has just lost most of her military weapons in a war with Israel. But wait a minute. Moscow comes to the rescue and rearms Syria beyond her prewar strength. Result? Syria moves closer to the Soviet orbit and becomes a belligerent Arab country.

Iran and Iraq need guns and tanks. How will they get them? To counterbalance United States influence in Iran, Moscow agrees to supply industrial and agricultural projects. In Iraq, Russia directly supplies arms. Result? Russia has a door cracked in Iran and opened in Iraq.

What about Egypt? Although Cairo has gone to the United States and France for arms aid, she is still dependent on the Soviets. No matter how edgy Cairo is for independence, she knows it would take years to retrain Egyptian forces to use alternate weapon systems. Some Russian advisors have left, but how many have stayed?

Although the Western world has made important strides in Middle East diplomacy, Russia continues to be a dominant force behind the Arab nations. She may keep a low profile at times, but she has many interests at stake, and she is always there.

"This conflict is a godsend for the Soviets," said a Western diplomat in Moscow about the Middle East's long no-war, no-peace situation. "If it didn't exist, they would have to invent it."[19]

Certainly Russia has made diplomatic, political, and military headway in the Middle East by encouraging unsettled conditions between Israel and the Arab countries. Ever since the United States rebuffed Egypt's President Nasser by refusing to sell him weapons in 1955 and withdrawing financial aid to build the Aswan High Dam, mammoth Soviet-financed development projects and sophisticated Soviet weapons have flowed freely into the Arab world.

The returns from this massive investment can be seen in striking ways. Thousands of military and civilian Soviet advisors are in Iraq and Syria. Many are in Egypt. Russian weapons are seen throughout the Arab world. Soviet cruisers bristling with missiles and advanced communications equipment regularly put in at Arab ports along the Mediterranean Sea and Indian Ocean.

Such an alignment of Russia with the Arab nations should not be surprising to those who read the Scriptures. Long ago the prophet Ezekiel warned the world of Russia's interest in the Middle East. God said that this northern power would come "against my people

86

of Israel, as a cloud to cover the land" (Ezekiel 38:16). Sometime before the return of Christ, Russia and her allies will move in quickly to take what they can from the oil-rich Middle East.

The present Russian interest in the riches of the Arab nations did not occur overnight. Over two hundred years ago, the Russian czars began weaving this sinuous web. Through the changing pattern of time, through dynasties, succeeding monarchies, revolutions, and upheavals, Moscow's persistent foreign policy thread runs through the Black Sea, the Mediterranean, and the Middle East.

Why is Russia interested in this part of the world? Throughout history, Palestine and Constantinople (now Istanbul) were the springboards from which the

At Masada, the Zealots committed mass suicide rather than submit to Roman armies. Photo courtesy of Israel Ministry of Tourism.

great powers sought to expand their empires. This region was the nerve center of communications between three continents, a land bridge between Asia, Africa, and Europe. Alexander the Great understood this and made it the goal of his Grecian Empire. The Caesars of Rome knew this and made it part of their empire. It is no wonder that Russia should also want this "golden jewel."

Another reason Russia needs this area is for self-preservation. All the Russian seaports are in the Baltic Sea, which is frozen most of the year. If Russia is to have a warm-water port, it must be on the Black Sea. This area is accessible only through the Bosporus and Dardanelles, the straits between Greece and Turkey. Russia needs to control the Black Sea, the Mediterranean, and the Middle East in order to control these straits.

Peter the Great (1682-1725) was the first to realize this. He transformed Russia from a relatively unimportant country into a major European power. He expanded the country westward toward Sweden and southward to Constantinople. Continuing this southward expansion was Catherine II (1762-1796), who fought two wars with Turkey and used the Muslim faith by tolerating and encouraging Islam to win friends in this area.

Through the years Russia has never had complete control over the Bosporus and Dardanelles straits, but she has had special treaties concerning them. Even today she is allowed to send warships through the straits by giving advance notice of such passage. But Moscow is not pleased with such an arrangement and has circumvented it by placing permanent notification on record with Turkey that Red Fleet vessels will pass through the straits.

Soviet intention to control the Middle East is clearly stated in Russian documents. The Ribbentrop-Molotov pact, negotiated and signed in November 1940, agreed that "the area south of Batum and Baku in the general direction of the Persian Gulf is recognized as the center of the aspirations of the Soviet Union." Is there any wonder why Soviet warships patrol the Mediterranean Sea today?

The prophet Ezekiel gives another reason for Russian expansion: the wealth of the Middle East. Russia will come against Israel "to take a spoil, and to take a prey" (Ezekiel 38:12). In that day people will ask Russia if she has come "to carry away silver and gold, to take away cattle and goods, to take a great spoil" (Ezekiel 38:13).

Throughout the centuries there has been wealth in the Middle East such as cedars in Lebanon, dates in Saudi Arabia, agriculture in Egypt. But why would a nation with the natural resources of Russia want these?

That was the question until the discovery of oil. All of a sudden, the Middle East took on an entirely new significance. Today the riches of the world are flowing into the Middle East because of the value and necessity of this precious commodity, oil. In a day when the need for energy is great and energy is unevenly distributed, the nations that control oil control the world.

This is what Defense Secretary Harold Brown has sensed. He declared, "In the event of some future confrontation the Soviet Union might be able to restrict access of the western world to its essential oil supplies." They could do this "by direct attack on the facilities of the major oil-loading ports, which lie near to Soviet territory." Could oil have been what Ezekiel meant when he wrote of a great spoil?

Although Russia is interested in the Arab states, Ezekiel says that one day she will be interested in Israel. Does this mean that Israel will have oil?

Some believe Moses may have predicted Israel would have oil. As he blessed the nation before his death, he said the tribe of Asher would "dip his foot in oil" (Deuteronomy 33:24). Although some take this to be general blessing, or olive oil, could Moses have been predicting oil from the ground, since the word "foot" is used? If this prophecy does refer to Israel having oil, it will be fulfilled when Christ is ruling the earth, a time when Israel will "offer sacrifices of righteousness" (Deuteronomy 33:19).

Much of the petroleum Israel did have was assigned to Egypt as part of the 1975 Peace Accord between Israel and Egypt. Moreover, geologists claim that the area west of the Jordan River does not have suitable conditions for important oil deposits because of a geological fault which lies parallel to the Jordan. This had resulted in the deep depression of the Red Sea, the Gulfs of Suez and Aqaba, the Dead Sea, the uplift-

The Dead Sea, which contains untold chemical wealth, is plainly visible from the Qumran Caves. Photo courtesy of the Israel Ministry of Tourism.

ing of the hills of Judea and Jordan, even up to western Syria and Lebanon. This does not mean, however, that the area is incapable of oil production. But since the land is broken rather than folded, the surviving structures capable of containing oil are believed to be small.

Israel has drilled for oil in the vicinity of Haifa, in the desert south of Beersheba, and in the Gaza vicinity.

The results, however, have been disappointing. Little oil has been discovered. A natural gas field was found, though, and an oil refinery has been built at Haifa. There is also a petrochemical plant in Israel.

Could oil be found in Israel in the future? There is the possibility of oil being discovered off the coast of Israel in the Mediterranean Sea. It is also possible that Israel may gain more land than she has now, some with oil fields.

There are other valuable products in Israel. For instance, there is great chemical wealth in the Dead Sea. This body of water is one of the richest chemical beds in the world. It contains bromine, sulfate, sodium, potassium, calcium, and magnesium. Moreover, its mineral salts amount to 24 percent of its volume, as compared with less than 4 percent in ordinary sea water. Israel has built entire cities in the desert to house the workers who are obtaining these chemicals from the Dead Sea.

Whatever the possibility for wealth, however, few recognized the economic worth of the Middle East until recently. Few, that is, except Ezekiel!

Another reason for Russia's interest in the Middle East is related to the goal of world-wide expansion of Communism. They believe that the enemies of Communism will surrender without resistance unless they are intolerably provoked. The way to spread Communism throughout the world is, therefore, to keep up a steady pressure and to encircle the NATO (North Atlantic Treaty Organization) powers.

By neutralizing and then winning over, one by one, the Afro-Asian states and then Latin American countries, Communism will be spread and the NATO countries surrounded. Communism has already spread throughout Eastern Europe. If Communism dominated the Middle East, Africa, and Latin America, the capitalistic countries would be surrounded. Marxism is already spreading in many of these areas; and the Middle East, the land bridge between Russia and Africa, is an important link in such expansion.

Moscow is also interested in the Middle East because if Russia commanded one or more of three strategic straits there, she could control much of the world's oil shipment. If Russia controlled the Strait of Hormuz at the inlet of the Arabian Gulf, she could block the oil

The Jordanian army, well equipped with modern tanks and trucks, is shown here parading before the royal stand at a military festival. Photo courtesy of Arab Information Center.

importation through that strait. If she controlled the Suez Canal, she could bottle up the oil that flows through that canal to Europe. Therefore, Egypt is very important to Russia. If Russia controlled the Bab e Mandeb Straits at the southern end of the Red Sea, she could control the supplies that move through there. She already has footholds in the Yemeni republics and could block the mouth of the Red Sea. What would happen if the Middle East oil supply to the world suddenly were cut off by one or more of these straits being closed? Without oil, how could the rest of the world fight Russia?

Will Russia be able to conquer the world? Not according to the prophet Ezekiel. It is true that she will come against Israel in the "latter days" before the return of Christ (Ezekiel 38:16), but a strange thing will happen. Instead of Israel defending herself against Russia, God will destroy Russia on the mountains of Palestine. God will rain upon Russia "an overflowing rain, and great hailstones, fire, and brimstone" (Ezekiel 38:22). This judgment will come directly from God. The Lord Himself will destroy this nation because of her greed and selfishness, and because she came against His chosen people.

We should realize that the Middle East is valuable to the major powers of the world as well as to Israel and the Arab nations. The world powers have important interests at stake and are deeply concerned with what happens in the Middle East. But God is sovereign and will one day bring to pass all that He has promised.

8

World War III

On a crowded street in Washington, D.C. stands an imposing building with large steel columns. Black limousines bring foreign dignitaries to its doors daily. Inside, its communications system can instantly reach the four corners of the earth. Across its desks come requests from the governments of many different nations.

This is the State Department of the United States Government, and the requests that are coming are for arms shipments to the countries of the Middle East. Each one is carefully examined, given consultation by the Pentagon, and then cleared for shipment.

The oil-producing countries of the Middle East are using much of their newfound wealth for military purposes. In 1955, Egypt signed arms agreements with Czechoslovakia. Since then, Egypt has bought from many countries, especially the Soviet Union and the United States. Syria has bought mainly from the Soviet Union. Iraq has bought from France, Britain, and the

Soviet Union. Saudi Arabia, Kuwait, Jordan, and Israel have bought mainly from the United States. Yes weapons are rolling through the floodgates of the Middle East. Rear Admiral Gene R. Larocque, once head of the Center for Defense Information, says that the weapons are sold all over the world, to any country that can afford them.

In a day of relative peace, it is difficult to consider the possibility of an all-out war. Yet we live in a world with many trouble spots. A national news weekly declares that wars occur with astonishing frequency and that in the last three decades there have been 119 armed conflicts involving sixty-nine nations!

The Middle East is receiving the most sophisticated weapons in the world, including tanks, missiles, ships,

Many Arab countries are now buying arms. Photo courtesy of Arab Information Center.

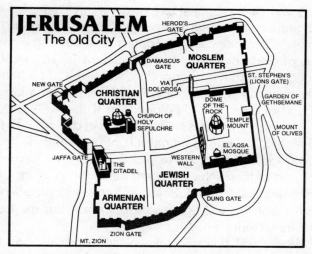

The plan of the Old City of Jerusalem is reproduced in this map. Courtesy of the Israel Ministry of Tourism.

planes, and electronic equipment. The Middle East countries want to protect themselves and their oil fields. But with so many arms going into that area of unrest, is it not likely that the weapons will be used?

Once again the Scriptures speak to a contemporary issue. Jesus spoke of coming war as a sign of the end of the age (Matthew 24:6). Other passages predict that one day great warfare will break out in the Middle East, warfare that will eventually usher in the return of Christ. Without doubt the Arab nations will be involved.

Some view this final world war as the battle of Armageddon. Although the place of Armageddon is involved, Scripture calls this conflict "the war of the great day of God, the Almighty" (Revelation 16:14, NASB). Notice, first, that it is termed a "war," not just a battle. Wars are made up of many battles and cover a longer period of time, as with World War II.

Also notice that it is called "the great day of God, the Almighty." Although men will fight one another, this is a war God allows to take place. When men come to the end of themselves, as they will in this conflict, God will bring it to an end. Satan and the Antichrist, along with mankind, will fight against heaven, but the Lord will be victorious.

Where will this war take place? Arab nations are afraid their land will be involved and have sought to protect their oil fields with the latest weaponry. The Scriptures predict that the land of Palestine will be the center of the conflict. The battlefield of Armageddon will be involved, for this is where the nations will be gathered (Revelation 16:16). This is a Hebrew name for the Valley of Megiddo in the north central part of Palestine, about halfway between the Sea of Galilee and the Mediterranean Sea.

Megiddo was historically an important military fort. Generals knew that if they wanted to control the northern part of Palestine, they had to command this valley. The fortress was also self-sufficient. Today a museum at Megiddo shows the underground springs that supplied water to this stronghold. The garrison also overlooks a large valley of which Napoleon said, "This would make the most natural place for the nations of the world to be gathered for a final war."

Another battlefield will be in the southern part of Palestine, in the land of Edom, or Idumea. God says that His "sword shall be bathed in heaven: behold, it shall come down upon Idumea, and upon the people of my curse, to judgment" (Isaiah 34:5).

The Valley of Jehoshaphat will be another place for this war. Some believe that this is located in central Palestine, between the city of Jerusalem and the Jordan River. This is where God will gather the nations and plead for His people (Joel 3:2). It is a place of judgment.

This will be a devastating time, a time of great judgment upon this world. Although we speak of a harvest of souls today, God says that a harvest of wrath is coming. An angel will "thrust in his sickle into the earth, and [gather] the vine of the earth, and cast it into the great winepress of the wrath of God" (Revelation 14:19).

There will also be a river of blood, perhaps 4-1/2 to 5 feet deep, right up to the horse bridles (Revelation 14:20). It will cover sixteen hundred furlongs, or two hundred miles, the length of Palestine today. What God is evidently confirming is that the whole land will be covered with war, from the north (the Valley of Megiddo), to the central (the Valley of Jehoshaphat), to the south (the land of Edom). This entire area will be covered with the blood of nations.

What countries will fight against Israel in this final world conflict? A careful study of Scripture shows that there will be four groups, or confederacies, of nations that will surround Israel in end-time. It is interesting how closely the present world scene parallels these end-time power blocs. Could it be that the nations today are aligned for this purpose?

One bloc of nations will be located west of Israel, in what is the area of the old Roman Empire. This bloc is mentioned in Daniel 2, the chapter that describes Nebuchadnezzar's dream of an image. Daniel later explains that this image portrays the whole sweep of Gentile history in terms of four world empires. Many Bible scholars agree that the fourth empire is Rome, which was at its zenith soon after the earthly life of Christ. We understand that the feet and toes of the image illustrate a reappearance, or continuation of Rome's great power.

The formation of a federation corresponding to the prophetic picture has been regarded as especially significant because of its central role in end-time events. Such an empire, Scripture indicates, will be the citadel of man's final struggle against God. It will give place to Christ's great Kingdom, a kingdom that will stand forever (Daniel 2:44-45). But until recent times no revived Roman Empire has appeared.

Some believe that the European Common Market may fit the description of this great power bloc. Begun on January 1, 1958, it was first an association of six nations—Belgium, France, West Germany, Italy,

"If I forget thee O Jerusalem, let my right hand forget her cunning" (Psalm 137:5). Photo courtesy of Israel Ministry of Tourism.

Common Market Building is located in Brussels, Belgium. Photo by Edgar James.

Luxembourg, and Holland. Fifteen years later, on January 1, 1973, the federation was enlarged by Great Britain, Ireland, and Denmark. The expressed goal of the association: the political union of Europe.

Admittedly, the areas occupied by the old Roman Empire and today's Common Market nations are not identical. But further shifts are yet possible. Within a short time, today's Common Market could reach its goal of a ten-member confederation with many associate nations. Together these nations might well occupy an area approximately the same as the old Roman Empire.

Some believe the Antichrist will rise from the revived Roman Empire. According to the Bible, his career will begin with the subjugation of three nations (Daniel

7:8), after which other countries will give their sovereignty to him (Revelation 17:13). He will make a treaty with Israel, guaranteeing her protection, but he will later break it and enter Israel for his own selfish ends. He will change laws to suit himself, persecute the people of God, and seek to be worshiped as God.

Many in Europe especially seem already to be looking for such a super leader. The mood is well expressed by Henri Spaak, one of the early planners of the Common Market. "We do not want another committee," he said; "we have too many already. What we want is a man of sufficient stature to hold the allegiance of all people and to lift us out of the economic morass into which we are sinking. Send us such a man and be he god or devil, we will receive him!"[20]

Another power bloc will come from the "north quarters" and "out of the north parts" (Ezekiel 38:6, 15). Because this bloc is said to come from north of Israel, many believe this confederation to be Russia, along with other allies.

This northern group will come against Israel "as a cloud to cover the land" (Ezekiel 38:16). The armies will come down to take the spoil from the land. They are going to make war against Israel.

How will this confederacy be destroyed? One would expect it would be by other nations fighting against it. However, instead of Israel fighting with this group of nations, God is going to destroy it. Fire, rain, hailstones, and brimstone are going to rain down from heaven, and God will bring this confederacy to an end (Ezekiel 38:22).

The third power bloc is east of Israel and called "the kings of the east" (Revelation 16:12). Little is known of this bloc since it is mentioned only twice in Scripture (Daniel 11:44; Revelation 16:12). The word "kings," however, is plural, so there may be more than one nation involved. Furthermore, it is east of Israel, so China, India, and other nations may be included. It is interesting to note the cleavage between China and Russia today, since there will be both a northern and an eastern confederacy involved in end-time events.

There is also a southern confederacy. This may be where the Arab nations fit in since many of them are south of Israel. The leader of these nations will be Egypt, called the "king of the south" (Daniel 11). Egypt, in spite of her lack of oil, does have political prestige. She has been called the leader of the Arab nations. This is the way it will be when this final war begins. It is when there is a balance of world power at the "time of the end" that Egypt will begin the great conflict (Daniel 11:40). She will push at Israel, causing the northern bloc of nations to come down against Israel.

Why will Egypt and Russia come against Israel? One reason is the "spoil of the land," which some have speculated may be oil.

Another reason Egypt will come against Israel may be the Arab's continuing hatred of the Jew. This is seen in the present underlying attitudes of these peoples as

Will Jerusalem be the final place of conten-
tion? Photo courtesy of Israel Ministry of
Tourism.

well as in the periodic outbreaks of terrorism around
the world. Speaking of the Arab nation of Edom, God
says she has had a "perpetual hatred" of Israel (Ezekiel
35:5). The psalmist also speaks of this hatred. He says
that the Arab nations have said, "Come, and let us cut
them off from being a nation; that the name of Israel
may be no more in remembrance" (Psalm 83:4). They
have also "consulted together with one consent"
(Psalm 83:5). Is this not happening in our world?

After Egypt pushes at Israel and Russia and her allies
come against Israel, God will destroy the northern armies in the mountains of Palestine. Until then, there
will be a balance of world power, but a miracle will
break that balance. Perhaps this is why the western
confederacy, the revived Roman Empire, will then

move into the "glorious land" (Daniel 11:41). This group of nations, led by the Antichrist, will take the treasures of the land, even the "precious things of Egypt" (Daniel 11:43). There is speculation there may be oil in the western part of Egypt to which this may refer. In any event, this is also the time when the Antichrist will put himself in the Temple and call himself god (2 Thessalonians 2:3).

When will God bring this war to an end? It is when the western confederacy, threatened by the kings of the east, comes back to the Holy Land to fight the eastern confederacy (Daniel 11:44). Instead of fighting each other, however, their animosity is turned against the Lord from heaven, who slays His enemies (Revelation 19:21). As Augustine said, the City of God will win, and Jesus Christ will be victorious when He comes again.

What lessons can we learn in light of these coming events? First, we should realize *the sovereignty of God*. God is sovereign and is working all things out for His purpose. Even heathen kings like Nebuchadnezzar of Babylon had to learn that "the most High ruleth in the kingdom of men, and giveth it to whomsoever he will, and setteth up over it the basest of men" (Daniel 4:17). God has a purpose for the nations of the world. Even today He is working all things out to bring glory to Himself.

Second, we should understand *the timing of God*. We need to sense the timing of God's program. Although the nations of the world may be staged for a final conflict, God is building His Church today, and we need to be involved in that program. God has commissioned each of us to make the Gospel available to all peoples. Some say the reason for meager missionary effort in Muslim lands is because many doors are closed. Perhaps a better assessment would be our degree of concern for these nations.

Third, we should have confidence in *the faithfulness of God*. As God has fulfilled His promises to nations in the past, so He will fulfill His promises to nations in the future. We can rest assured that He will keep His promises to us. God will accomplish all He has promised, and the believer today is in the very center of His concern.

9

When God Rules the Nations

After years of struggle, months of intense negotiation, and several major wars, the question still remains: When will there be real, lasting peace in the Middle East? I asked this question of an Israeli army captain. He said, "Peace? What peace? We have always lived with war. But we must give peace a chance. There are no guarantees, no assurances. You ask me, 'Can Israel trust the Arabs?' It will take years!"

An optimistic Arab businessman from East Jerusalem said he felt recent initiatives were good.

"But will they bring real peace?" I asked.

"Oh no!" he said. "You'll never have real peace with a piece of paper. The only way there will be real peace in this part of the world is that either the Arabs will defeat all the Israelis or the Israelis will defeat all the Arabs. Without God ruling the nations, this is the only way there will be real peace."

God ruling the nations? This is exactly what the prophets predict. One day God will change hearts, transform military weapons into instruments of peace, and rule the nations with a rod of iron. He says, "Nation shall not lift up sword against nation, neither shall they learn war any more" (Isaiah 2:4).

Every generation has dreamed of a time when there would be no war and a place where everything would be right. In 1516 Sir Thomas More wrote of Utopia, an ideal commonwealth whose inhabitants exist under perfect conditions. Long before him, Plato, in his *Republic,* mentioned a similar idea. Even Arab legends describe an earthly paradise in the western, or Atlantic Ocean.

Rather than righteousness prevailing over a small, confined place on the earth, the Scriptures predict that righteousness will reign over the entire earth. One day there will be a whole new world order, a new economic system, a new administration of justice, and peace. God says that He will give Christ "the uttermost parts of the earth" for His possession (Psalm 2:8).

Throughout history there has been a constant struggle of man against God. Augustine pictured this struggle as the City of God against the City of Evil, with the City of God eventually triumphing. The psalmist speaks of the kings and rulers of the earth counseling among themselves and rebelling against God. He says that they are "against the LORD, and against his anointed" (Psalm 2:2). They want to break away their bands, break away their cords, have their freedom.

But this is not God's way, this is not God's plan. Instead, He will set His "king upon my holy hill of

The Golden Gate in the wall of Jerusalem can be seen from the Garden of Gethsemane. Some believe this gate will remain shut until the return of Christ in fulfillment of Ezekiel 44:1-3. Photo by Edgar James.

Zion" (Psalm 2:6). One day, when Christ comes again, He will break the nations "with a rod of iron; thou shalt dash them in pieces like a potter's vessel" (Psalm 2:9). This was Augustine's victory of the City of God.

Such an event must await the time when Christ Himself returns to the earth as was promised when He ascended. He is to come "in like manner as ye have seen him go into heaven" (Acts 1:11). This, of course, means that He will be visible and will personally reign upon the earth. His reign will be one of peace and righteousness (Isaiah 9:6; Hebrews 1:8).

Will He rule the nations? Without question there will be whole nations in existence when Christ reigns on earth. John says that the "nations of them which are saved shall walk in the light of it" (Revelation 21:24). The Scriptures predict a glorious future for Israel and her everlasting possession of the land. The Bible also foretells a future for some of the Arab countries, such as Egypt and Assyria.

The kind of reign Messiah will have is spoken of in two identical Old Testament passages (Isaiah 2:2-4; Micah 4:1-3). The fact that this information is repeated exactly shows its importance. All the characteristics of Christ's future reign contrast those of present human governments. What are the characteristics of Christ's reign mentioned in these passages?

The reign of Messiah will be exalted. The prophets speak of Christ's reign being "established in the top of the mountains" (Isaiah 2:2) and "exalted above the hills" (Isaiah 2:2). This is a goal many nations have purposed, but none has reached. Babylon, Medo-Persia, Greece, and Rome all wanted to be world empires. Russia wants to dominate the world and bring all peoples under the communistic social structure. The Arab nations want the world to notice their existence and problems. When Christ returns, His Kingdom will be exalted above all others. Christ will be King among all kings and Lord among all lords (Revelation 19:16). Because He will be exalted, His Kingdom will be exalted. The exaltation of His Kingdom is seen also in the other characteristics.

The reign of Messiah will be universal. The Scriptures predict that "all nations shall flow unto it." Although there will be other nations during this time, Messiah's Kingdom will dominate all of them.

During Messiah's reign, God will fulfill the promise

111

made to Israel. She will have the land for an everlasting possession. Although the land from "the river of Egypt unto the great river, the river Euphrates" was promised to her, Israel has never yet possessed all of this land. Even when Solomon reigned, Israel did not possess all the land, because Solomon ruled through puppet rulers who paid tribute to him (1 Kings 4:21). Since God is a God of His Word, Israel must possess the land as an everlasting possession.

Israel must also have an everlasting kingdom, since her kingdom is to be established forever (2 Samuel 7:16). This means that Israel will not go out of existence since she as a nation is promised to be forever. Although other nations have come and gone throughout history, Israel is promised to remain forever.

Israel must also have an everlasting king, since the house and throne are also to be established forever (2 Samuel 7:16). The only One who could fulfill such a promise is the Messiah. This is why the angel promised Mary that Christ would have the "throne of his father David" and that of "his kingdom there shall be no end" (Luke 1:32, 33).

As previously mentioned, Arab nations will be in existence during this time. The prophets predicted that Egypt will know the Lord in that day, and that God will heal that land (Isaiah 19:21-22). There will be a place for Assyria, too, since the Egyptians will serve with the Assyrians (Isaiah 19:23), and a highway will be built between Egypt and Assyria.

Messiah will rule over all these nations, and they will

do His will. Daniel shows this when he speaks of a great Stone that smites the feet of the image of Gentile world power (Daniel 2:35). The Stone, of course, is Christ who will smite Gentile world power at His second coming. The Stone will then become a great mountain, a great kingdom that will fill the whole earth. Messiah's Kingdom will dominate all others, for He will rule the nations "with a rod of iron" (Psalm 2:9). His rule will be universal.

The reign of Messiah will be righteous. The prophets point out, that "He will teach us of his ways, and we will walk in his paths" (Isaiah 2:3; Micah 4:2). Since the King is righteous, His ways and paths must be righteous. Although the most efficient form of government is a dictatorship, the problem with such a government is the unrighteousness of the dictator. This will not be true with Messiah's reign. Since He is righteous, His rule will be righteous, and He will require uprightness of His subjects.

The righteousness of God is set forth in the law, or revelation God will give from Jerusalem. It is the standard that His subjects will obey. No one will be ignorant of the law, since the people will be taught by the Lord. They will obey the Word of God and walk in His paths. The whole age will be characterized by righteousness—righteousness taught by the Lord and exhibited in His people.

The reign of Messiah will be just. Righteousness is the application of God's standard, whereas justice is the meeting of that standard. The prophets affirm that the Lord will judge among the nations (Isaiah 2:4; Micah 4:3). God will deal with those who do not keep His Word. His reign will be one in which true justice prevails.

A major judgment of the nations, including the Arab nations, will occur at the beginning of Messiah's reign. When He returns to the earth, before Him will be "gathered all nations: and he shall separate them one from another, as a shepherd divideth his sheep from the goats" (Matthew 25:32).

The word translated "nations" is the Greek word *ethne*. (From the same word we get *ethics.)* It is the same word used in the Great Commission, in which we are told to disciple all nations (Matthew 28:19). The people who are alive at the earthly return of Christ will be separated the one from the other according to

God will one day separate the wheat from the chaff as is done in the Middle East today. Photo by Edgar James.

God's standard. Those who are righteous and have kept His Word will go into the Kingdom (Matthew 25:34). Those who have done evil and have not obeyed Him will be one day put into everlasting fire (Matthew 25:41). Although national groups are involved, this will be an individual judgment.

On what basis will these people be separated the one from the other? Jesus said it is because some fed Him, gave Him to drink, and took Him in (Matthew 25:35) when they had done these things to "the least of these my brethren" (Matthew 25:40). One's works always indicate his faith, and this will be especially true in the time of trouble when the nations war against Israel. The Antichrist will lead this persecution of many believing Jews. The persecution will be so severe that some Jews will flee to the mountains (Matthew 24:16). When people have an opportunity to help Christ's brethren after the flesh, the saved Jews, some people will take the opportunity, and some will not. Such a work of faith will demonstrate the person's own faith in God, so it is on this basis that people will be separated. God's principle still is true, "I will bless them that bless thee, and curse him that curseth thee" (Genesis 12:3).

In which group will the Arab nations be when people are judged? Those Arabs who have accepted Christ's salvation will be separated for His kingdom, but the unbelievers will be cut off because Messiah's rule will be a just one.

Justice will prevail during Messiah's reign. Although people will be born sinners, those who commit gross

115

acts of sin, such as murder, will be cut off. The Lord will righteously judge all people because righteousness will characterize this age. The sinner, even though a hundred years old, "shall be accursed" (Isaiah 65:20).

The reign of Messiah will be one of peace. War is the result of hostility between people or nations. During an age of righteousness and justice such as Christ's reign will be, how can there be any war? The conflicts and struggles against injustice will no longer be necessary. All will obey the Lord and walk in His paths. All will be at peace with the Lord and with one another.

Because Messiah's reign will be peaceful, the instruments of war will become instruments of peace. Swords will be turned into plowshares and spears into pruninghooks. The nations will not "learn war any more" (Isaiah 2:4; Micah 4:3).

Why doesn't God rule the nations today? They have not accepted His ways and His provision. They have put themselves first and have rebelled against Him.

This is what happened at the beginning, when God made man. Everything Adam had was good, and there was peace in the garden. But Adam did not want to follow God's ways. Adam rebelled and disobeyed God and usurped God's authority. Consequently Adam and Eve were expelled from the garden, and man has suffered ever since.

Israel did not want God's rule in her life either. God picked this nation to be an object lesson to others so they could see a demonstration of His love, but Israel wanted a king as the other nations. She did not want to be ruled by God. So the Lord gave them a king, first Saul, then David, and then Solomon, but the people continued to disobey God.

Other nations have disobeyed as well. Rather than seek the Lord and be ruled by Him, they have turned to their own ways, they have followed their own paths. This is why there is a struggle around the world and in the Middle East.

Although the changing of the hearts of nations awaits Christ's return, He can and does change the hearts of individuals today. Jesus promises this to all who accept Him. He says, "Peace I leave with you, my peace I give unto you" (John 14:27). Though man has turned his back on God, God can forgive man's sins. In Christ's death, He made provision for peace with God. We receive this by accepting Christ's provision for us.

A man's heart must be changed before a man's life can be changed. The man who accepts Christ as his Saviour attains also the peace *of* God in the midst of problems because God promises to care for each of us.

When will we have real, lasting peace? The nations of the world will not have this kind of peace until Jesus comes again, but you and I can have peace today, both the peace with God and the peace of God.

10

How to Approach an Arab Today

In London's major newspaper, the *Daily Express,*
Saudi Arabia advertises for barge masters and cleans-
ing managers and offers large, tax-free salaries; free
accommodations; free food and transportation; and
three weeks leave for every three months of work. The
United Arab Emirates asks for college teachers and
adds a car allowance to the same generous benefits.
The Bechtel Corporation, an American company in
San Francisco, needs to import workers to construct an
industrial city for 200,000 people on the Arabian Gulf.

Through the years there has been only minimal contact with Arab peoples because they kept mostly to
themselves. But now these countries that once were
closed to the West are crying for help. In a peninsula
half the size of the continental United States, there are
only five to ten million people, making a dearth of
labor. Moreover, natives of the oil-rich countries prefer
steady jobs in armies or governments, and the

118

nomadic Bedouins want to raise camels. Hordes of workers must be imported. Development plans show that, for the next five years, building alone will require at least a doubling of the one million foreign workers already present in Saudi Arabia.

Arabs, because of their wealth, also travel to other nations. Some own homes in London or Paris, while others have businesses in Canada or the United States. Recent estimates show there are over one million Arabs in Europe alone, working in factories and doing many different jobs. On a recent visit to London I witnessed Arabs working as shoe and suit salesmen, waiters, and museum tour directors. There are Arabs in the airports and on the streets. Saudi Airlines has one of the nicest travel offices in London.

Through the discovery of oil, Arabs have been ushered into the twentieth century. They have contact with world ideas, world thought, world politics. They are receiving visitors to their own lands and traveling to many others. What are they learning from today's world, and what are they learning from today's faith?

In the past it was difficult to visit Arab lands; today it is easier. Some may travel to Arab nations or meet Arabs in other countries. Others may know people who are going to visit Arab countries. It is also possible to meet an Arab at a convention, in a store, or on a bus.

How does one approach an Arab when one meets him? Is there any special way to get to know an Arab? And how does one share his faith with an Arab?

Arabs need to know God loves and cares for them.
Photo by Chuck Wagner.

Arabs are people just like anyone else, but they have lived in relative isolation. Most are very religious people, bound to one another by the Muslim religion. Second only to language and culture, religion plays a very important part in the life of an Arab. For the most part, however, it is a religion of doing rather than believing. It is marked by prayers, pilgrimages, and propagation.

In approaching an Arab today, I would suggest the following principles to make our witness effective.

1. *To be an effective witness to Arabs, we must meet their needs.* Jesus dealt with people in different ways because the people were different. They had different backgrounds, different views, different circumstances. Of course, the message of the Gospel was the same in each case, but the way it was presented was different.

For instance, have you ever compared the way Jesus dealt with Nicodemus to the way He dealt with the woman at the well? He appealed to the differences between these two in need. Nicodemus was a man, she was a woman. He was a Jew, she was a Samaritan. He was learned, she was ignorant. He was moral, she was immoral.

Jesus used two approaches because of these differences. He appealed to Nicodemus' knowledge, to his intellect as a teacher. He asked, "Art thou *the* teacher of Israel, and understandest not these things?" (John 3:10, ASV, italics added). To the woman, Jesus appealed with tact and persistence. He won her friendship by asking for a drink (John 4:7). It was the same Gospel, but different approaches.

In the same way, dealing with Arabs is different from dealing with other people. Arabs are firm believers in hospitality and friendship, and these traits come before anything else. Most Arabs are also believers in Islam and should be approached differently from the way one approaches a godless person. Arabs are also much more interested in the spoken word than in the written word. Stories play an important part in the lives of Arabs. They have a different history and culture and should be appealed to on the basis of it. But it is still the same Gospel and the same message, and they need the same Saviour.

2. *To be an effective witness to Arabs, we must deal with the individual.* Although there may be opportunities to speak to Arabs in groups, it is not wise to talk to a Muslim about his faith in front of others. He will feel compelled to defend his faith or promote his religion. This is why it is important to make personal contacts.

It is easy to open a conversation about God with an Arab. The following suggestions may prove helpful in such a discussion.

a. An Arab would be interested in the resurrection of the dead. When there is a death in a family, not only should sympathy be expressed, but it may be an opportunity to discuss how Jesus will one day raise the dead. Resurrection is of great interest to Muslims because they look to a day of future judgment. Important

Will this child know the peace God can give? Photo by Chuck Wagner.

passages for them to consider are John 5:28-29 and John 11:1-44.

b. An Arab would want to know about the forgiveness of sin. Although Muslims have a consciousness of sin and seek God's forgiveness, there is no assurance of such forgiveness. It is helpful to show them that Jesus has the power to forgive sin (Luke 5:18-26), and that David had assurance he was forgiven (Romans 4:6-8). God says that we can "know that ye have eternal life" (1 John 5:13).

c. Explain about the Lord Jesus Christ. Muslims believe in Christ, but only as a good man or teacher. Tell the Muslim that Jesus died for man's sins and that Jesus can transform a man's whole life. Important passages are Luke 19:1-10; Matthew 1:21; and Luke 2:11.

3. *To be an effective witness to Arabs, we must remember that many are believers in God.* Their cry is "There is no God but Allah," and this expression is ingrained in their hearts. They do have a sense of God's judgment. They are aware of their shortcomings and failures, but their attitude is one of a slave before his master. When they do wrong, they only hope for forgiveness from God. It is important to show that God can and does forgive sin. God gives a full and free pardon for sin when we accept God's provision in the death of Christ.

4. *To be an effective witness to Arabs, we must remember that the message is judged by the character of the messenger.* Believers are called "ambassadors for Christ" (2 Corinthians 5:20), and they are such when dealing with the Arab people. Some Arabs feel that throughout history, Western people have approached them in a hostile manner. As an ambassador for Christ, it is therefore important to show a practical demonstration of love. This may include hospitality, honesty, faithfulness to one's word, or any action in which they can see one's sincere purpose. They want to see the love of Christ demonstrated to them, a Christianity that works.

5. *To be an effective witness to Arabs, we must remember the importance of the Bible.* Although a message may be forgotten, the Scriptures speak continually. When one recognizes that the Scriptures are the Word of God, one reveres them.

It is important to treat the Bible with the same respect that Muslims treat the Koran. They never hold the Koran lower than the waistline, for they dare not trample it under foot. In the same way, the Bible, any portion of it, or tracts, should be held in reverence.

Some believe in having Muslims pay something for the Scriptures rather than just giving them the Bible. Muslims may appreciate it more when they have to pay for it. However, this depends on one's attitude toward this idea and one's friendship with the Arab. In any event, an Arab will read the Bible if it is honored as the Word of God.

6. *To be an effective witness to Arabs, we must remember that they respond to love.* This is true with any people, but it is especially true of Arab people. Arabs believe in the morality of friendship. Before one can discuss God with an Arab, there must be friendship and hospitality. One must accept the Arab as a person.

Many have failed through the years as they have tried to battle Islam with argument, abuse, and ridicule. But practical love will break down many barriers. It is important to smile and greet Arabs with affection. Share the details of your life with them. Invite them to your home. Respect them, and show honor to them. Love them by trying to understand their point of view. Sit with them to read. Respect their history and culture. Before an Arab trusts your Lord, he must first trust you.

7. *To be an effective witness to Arabs, we must remember the work of the Holy Spirit.* We must pray for the work of the Holy Spirit to convict of sin and

bring salvation. One can speak the message and love others in the Lord, but it is the Spirit of God who saves. The messenger of God must be dependent on the Lord. Arabs are being saved today, and we pray that many more may come to know the Lord in these strategic times.

8. *To be an effective witness to Arabs, we must have a faith to share.* The world may be divided into two major

A young Arab boy in Jerusalem carries trays through the alleys of the old walled city. Photo courtesy of the Israel Ministry of Tourism.

groups of people, those who are believers and those who are doers. The doers think they must earn their way to heaven. But God has shown us that no one can buy his way to heaven. The prophet Isaiah said that "all our righteousnesses are as filthy rags" (Isaiah 64:6). There is nothing we are capable of doing to earn our way to heaven.

How can a sinful person have fellowship with a holy God? Jesus gave the answer. He said, "I am the way, the truth, and the life: no man cometh unto the Father, but by me" (John 14:6). Jesus, the perfect Son of God, died for the sins of the world. He died for your sins and mine and paid the penalty that should have been ours. Simply by accepting His provision, we can have eternal life; we can go to heaven. The apostle Paul said, "Believe on the Lord Jesus Christ, and thou shalt be saved" (Acts 16:31). Jesus wants to deliver us from sin, to give us His righteousness, to give us a new life in Him. Have you trusted Him for that? And if not, won't you now?

When you trust Christ for your salvation, then you can share this wonderful faith with others. Maybe one day you will be able to share your faith with an Arab.

Notes

1. Ray Vicker, *The Kingdom of Oil* (New York: Scribner's, 1974), p. 20.
2. "Oil: A Glut—And a 'Crisis,'" *Newsweek,* May 24, 1976, p. 69.
3. April 18, 1977.
4. "All About the New Oil Money," *Newsweek,* February 10, 1975, p. 59.
5. Ibid., p. 60.
6. Thomas J. Abercrombie, "The Sword and the Sermon," *National Geographic,* July 1972, p. 33.
7. Edward G. Browne, *Arabian Medicine* (Cambridge: Cambridge U., 1921), p. 44.
8. Thomas J. Abercrombie, "The Sword and the Sermon," *National Geographic,* July 1972, p. 43.
9. Ray Vicker, *The Kingdom of Oil,* p. 70.
10. John J. Putman, "The Arab World, Inc.," *National Geographic,* October 1975, pp. 504, 507.
11. "Why Peace Doesn't Buy Prosperity for Egypt," *U.S. News & World Report,* August 23, 1976, p. 58.
12. Ray Vicker, p. 55.
13. Ibid., p. 163.
14. Ibid., pp. 211-12.
15. Ibid., p. 212.
16. "Jews Participate in Arab Symposium," *Midwest Arab Perspective,* November 1976, p. 4.
17. Philip Gillon, "The Core of the Problem," *The Jerusalem Post Magazine,* September 12, 1975, p. 9.
18. "How Three Heroes See It," *Newsweek,* May 24, 1976, p. 38; cf. *Chicago Tribune,* September 4, 1977.
19. "Bear Hugs and Kalashnikovs," *Time,* June 9, 1975, p. 26.
20. Edgar C. James, "Prophecy and the Common Market," *Moody Monthly,* March 1974, p. 44.